Practical Digital Electronics for Technicians

Will Kimber

Second edition

Newnes
An imprint of Butterworth-Heinemann
Linacre House, Jordan Hill, Oxford OX2 8DP
225 Wildwood Avenue, Woburn, MA 01801-2041
A division of Reed Educational and Professional Publishing Ltd

 A member of the Reed Elsevier plc group

OXFORD BOSTON JOHANNESBURG
MELBOURNE NEW DELHI SINGAPORE

First published 1994
Second edition 1997

British Library Cataloguing in Publication Data
Kimber, Will
Practical digital electronics for technicians. – 2nd ed.
1. Digital electronics. 2. Digital electronics – Problems, exercises, etc.
I. Title
621.3′815

ISBN 0 7506 3750 1

Typeset by Laser Words, Madras, India
Printed and bound in Great Britain by MPG Books Ltd, Bodmin, Cornwall

Contents

Preface ix

List of practical exercises xi

1 Setting the scene **1**
Introduction 1
Analog and digital signals 2
The practical exercises 3
The pulse 4
Circuit diagrams 5

2 Logic gates **7**
Types of logic gate 7
The AND gate 7
The OR gate 9
The NOT gate 10
The NAND gate 10
The NOR gate 11
Exclusive OR and exclusive NOR gates 12
Logic gate packages 14
Multiple input logic gates 19
Logic rules and theorems 21
Simplification of logic expressions 25

3 Combinational logic **28**
Problem solving 29
Derivation of the logic statement from the truth table 32
The use of pulses as input signals 34
The traffic light system 37

4 Karnaugh mapping **40**
The Karnaugh Map (pronounced 'car-no') 41
The two-variable Karnaugh map 42
A simplification 45

The three-variable Karnaugh map 48
The four-variable Karnaugh map 51
Don't care/can't happen states 53
Static hazards 57

5 Logic families 62
The three types of family 62
TTL – transistor transistor logic 62
CMOS – complementary metal oxide semiconductor 63
ECL – emitter coupled logic 63
Characteristics of logic families 63
Supply voltage 64
Logic levels 64
Propagation delay 66
Fan-in and fan-out 68
Noise margin 70
Power dissipation 72
Speed–power product 72
TTL logic 73
The basic TTL circuit 73
The standard TTL NAND gate with totem pole output 74
The open collector gate 78
The wired AND/OR gate 80
Sourcing and sinking 82
The pull-up resistor 85
Tristate logic devices 87
CMOS Logic 88
The CMOS NOT gate (inverter) 89
The CMOS NOR gate 89
The CMOS NAND gate 91
The ECL OR/NOR gate 91
Interfacing 93
Interfacing CMOS to TTL 93
Interfacing TTL to CMOS 95
The Schottky diode 95
Schottky TTL 95

6 Sequential systems 1: Flipflops 97
The bistable condition 97
The latch 99
Bistable (or flipflop) circuits 100
The RS (reset–set) bistable or flipflop 101
The clocked RS flipflop 106
The D-type flipflop 110
The latch 114
Single-bit memory with write and read facility 115

The JK flipflop 119
The master–slave JK flipflop 123

7 Sequential systems 2: Counters and shift registers **128**
Counters 129
The 3-bit asynchronous counter 129
The 3-bit synchronous counter 136
The 4-bit synchronous counter 138
Shift registers 143
The serial load shift register 144
PISO and PIPO shift registers 150
The universal shift register 152

8 Schmitt triggers and multivibrators **158**
The Schmitt trigger 158
Multivibrators 164
Main categories of multivibrator 165
The 555 timer ic 176
Logic gate oscillators 177

9 MSI combinational logic circuits **180**
Multiplexers 180
Demultiplexers 186
The data transmission system 188
Encoders 190
Decoders 194
Code converters 197
Adders 198
The 4-bit parallel adder 201

10 Display devices **207**
The light emitting diode (LED) 207
LEDs as indicators 210
The seven-segment display 210
The decoder/driver 213
The seven-segment display with counter and latch 216
The liquid crystal display (LCD) 219
The dot matrix display 219
A final word on LED devices 221

11 Analog and digital conversion **222**
Digital-to-analog converters (DACs) 223
The binary weighted type 223
The *R*-2*R* ladder network 226

Analog-to-digital converters (ADCs) 227
The counter-ramp type of ADC 227
Alternative types of ADC 230

12 Fault finding **231**
Analog fault finding compared with digital fault finding 231
Typical faults 231
The procedures 232
Some examples of fault conditions 234
Equipment for digital fault finding 239
The logic probe 239
The logic pulser (pulse generator) 240
The logic monitor (logic clip, logic checker) 240
The current tracer (current checker) 241
The logic analyser 241

Answers to questions 242

Index 259

Preface

This book combines an explanation of the basic theory of digital techniques with a total of forty-two practical exercises, all of which can be carried out with equipment that is normally readily available in departments running electronics courses.

These exercises, for which 'cookbook'-type instructions are given, are intended to provide the opportunity to become familiar with a wide range of digital integrated circuits and gain a real grasp of the way digital systems work. They are based on my own experience of teaching electronics at all levels of further education.

Throughout the book there are also many questions (answers provided, where appropriate), some of which require further research and reading. A knowledge of electrical principles and electronics to BTEC Level II, or its equivalent, is assumed.

The book is intended to cater for students following Advanced GNVQ, BTEC NII/NIII, GCE A-level and City & Guilds Courses. Furthermore, I hope it will prove a helpful text for practising electrical and mechanical engineers who intend to get to grips with the world of digital electronics.

Will Kimber

Practical exercises

2.1 To become familiar with the five basic logic gates 14
2.2 To investigate the use of the universal NAND and NOR gates 17
2.3 To verify De Morgan's theorems 24
3.1 To verify some examples of combinational logic 32
3.2 To investigate logic circuits with the use of pulse input waveforms 35
3.3 Traffic lights (1) 38
4.1 Traffic lights (2) 56
5.1 To investigate fan-out 68
5.2 To investigate logic levels 71
5.3 To investigate the voltage levels of a TTL NAND gate 76
5.4 To investigate the wired AND gate 81
6.1 The two-transistor switch 97
6.2 To investigate the action of the RS flipflop 101
6.3 To investigate the action of the D flipflop 111
6.4 To investigate a single-bit memory unit 117
6.5 To investigate the action of the JK flipflop 119
6.6 To investigate the action of the Master–Slave flipflop 125
7.1 To investigate the 3-bit asynchronous binary counter 131
7.2 To investigate the reset action of the 4-bit asynchronous binary counter 134
7.3 To investigate the action of the 3-bit synchronous counter 137
7.4 To investigate the 4-bit synchronous up counter 139
7.5 To investigate the 4-bit synchronous up–down counter 141
7.6 To investigate the 4-bit serial load shift register using D flipflops 145
7.7 To investigate the 4-bit parallel load shift register 150
7.8 To investigate the universal shift register 152
8.1 To investigate the Schmitt trigger using (a) the 741 operational amplifier ic, (b) the 7414 Schmitt trigger ic 159
8.2 To investigate the astable multivibrator using (a) the 741 operational amplifier ic, (b) the 555 timer ic, (c) the 7414 Schmitt trigger ic 166

8.3 To investigate the monostable multivibrator using (a) the 555 timer ic, (b) the 74121 monostable ic 172
8.4 To investigate the crystal-controlled oscillator 178
9.1 To investigate the multiplexer 182
9.2 To use the multiplexer to implement a Boolean function 185
9.3 To investigate the demultiplexer 187
9.4 To set up a data transmission system 188
9.5 To investigate the encoder 192
9.6 To investigate the decoder 196
9.7 To investigate the full adder 204
10.1 To investigate the LED 208
10.2 To investigate the seven-segment display 212
10.3 To investigate the decoder/driver with seven-segment display 214
10.4 To investigate the seven-segment display with counter and latch 216
11.1 To investigate the DAC 224
11.2 To make up and test an ADC 228

1 Setting the scene

Introduction

The term **digital** refers to a system in which the signal voltage can have only two values, provided by a basic switching action. **Electronics** is concerned with the movement of electrons in a vacuum, gas or semiconductor.

Digital electronics therefore could have been developed from the early 1900s, following the invention of the thermionic valve. In fact it was not until 1945 or thereabouts that the world's first computer was built. This weighed in at 30 tons, contained 18 000 thermionic valves and 6000 relays and had a power rating of 150 kW. The main obstacle to progress in this particular area was the sheer physical size of the equipment. Today's personal computers do not weigh 30 tons ...

The capabilities of modern computers have been entirely dependent on advances in the understanding of semiconductor physics, design capability and, not least, developments in manufacturing technology.

Microelectronics is defined as 'the design, manufacture and use of electronic units using ... solid-state components, especially ... integrated circuits'. The age of microelectronics started as comparatively recently as 1948–49 with the development of the transistor by Bardeen, Brattain and Shockley. The speed of progress since then hardly needs emphasizing.

Integrated circuits (ic's) were produced in the early 1960s, at high cost. They contained about 24 discrete components (resistors, capacitors etc.) and were for analog rather than digital applications. In today's 'digital world' the component density is somewhat higher, at about 10 000 for some applications.

The popularity of digital integrated circuits compared with their discrete component equivalents is due to a number of important factors:

- A wide range of devices available
- The fabricating process produces circuits that are more complex and are physically smaller and lighter
- Replacement is not difficult, with the correct desoldering tool, even for surface mounted components
- They are more reliable, both mechanically and electrically
- They offer enhanced performance (less noise and distortion, for example, in CD systems)

Analog and digital signals

An **analog** signal is one that varies continuously with time between two limiting values. It can have any value whatsoever between these two (maximum and minimum) limits. See Figure 1.1.

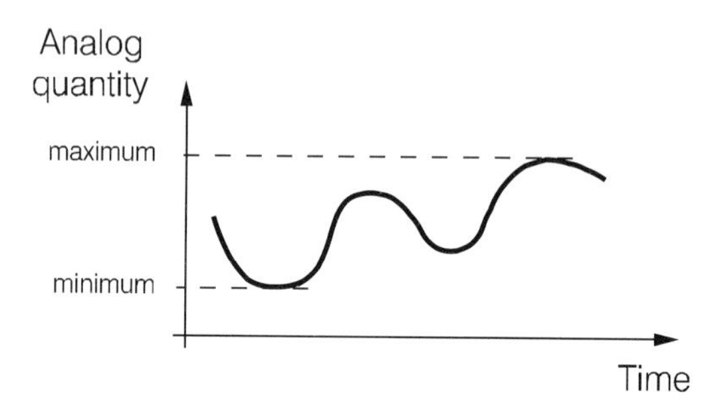

Figure 1.1 *The analog signal*

Examples of devices using analog signals include:

- (The pointer of) a moving coil meter
- An alcohol (or mercury) thermometer
- A car temperature gauge

These are all measuring physical (that is non-electrical) quantities. In order to be useful in our system the physical measurement must be converted into an appropriate electrical signal, which would normally be a voltage. This electrical signal is known as the **analog** of the physical quantity.

Analog signals are susceptible to noise and interference.

A **digital** signal is one that is expressed in terms of discrete (distinct) fixed levels. In sampling techniques, any number of levels may be used; in logic systems there are just two levels, for example +5V(**high** or **logic 1**) and 0 V (**low** or **logic 0**). Such a system can be regarded as being either 'on' (logic 1) or 'off' (logic 0). See Figure 1.2.

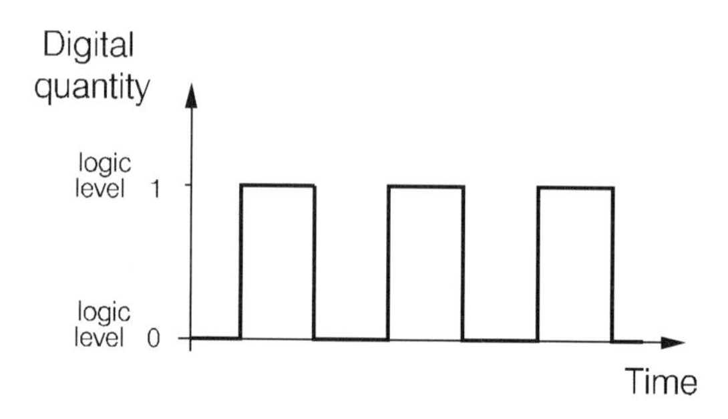

Figure 1.2 *The digital signal*

Examples of devices using digital signals include:

- A light switch
- A buzzer
- A burglar alarm

Questions

1.1 State whether the following signals are analog or digital.

(a) The indication of speed of a car along a winding road
(b) A pocket calculator
(c) A hi-fi amplifier
(d) A sine wave oscillator
(e) The emf from a thermocouple
(f) The gate count at a football match
(g) The human voice

1.2 Digital displays feature nowadays in such applications as clocks, petrol pumps and supermarket checkouts. What do you consider to be the advantages of digital displays? Are there any disadvantages?

The practical exercises

Typical equipment and the components required are listed for each exercise. The main points to be stressed here are that the +5 V supply for the digital ics should be well regulated (their specification calls for a tolerance of ±0.25 V), and decoupled with a (parallel) 10 μF tantalum capacitor at board entry. In addition, it is advisable to connect a 0.01 μF (10 nF) disc ceramic capacitor across the supply pins of the ic.

The choice of ic is between TTL and CMOS. CMOS are lower power devices than TTL but are prone to damage from static electricity from the human body, although they do have a measure of internal protection. Since, at any one time, the practical exercises use small numbers of TTL devices, the resultant current demand is unlikely to be beyond even the most basic of power supplies. As may be already known, even within the TTL 74 series family, there is a choice, ranging from low speed–low power to high speed–high power devices.

With this in mind, in the interests of standardization and coupled with the possible static problem of CMOS (and a small amount of personal preference), the TTL LS type device is suggested, where available, for all exercises. It is safe to predict that these exercises can be performed with success using any type of TTL 74 series ic.

The manner of assembly of a particular circuit is, of course, a matter of choice for the practitioner. Ready-made units are available, at a price, but part of the learning process surely lies in assembling all the bits and pieces and gaining more experience from so doing. Any extra time involved brings its own reward! Several breadboard-type systems are available where component wires push into holes on a 0.1 inch grid and are held in place by spring contacts. Among the firms marketing such boards are Maplin (retail) and RS Components (trade).

The pulse

A number of exercises require a so-called **clock pulse** as an input to the system. This pulse is merely a change in logic level (from 0 to 1 and back to 0) and could be achieved by connecting the input from 0 V to +5 V and back to 0 V (see Figure 1.3), using a toggle or push-button switch or similar.

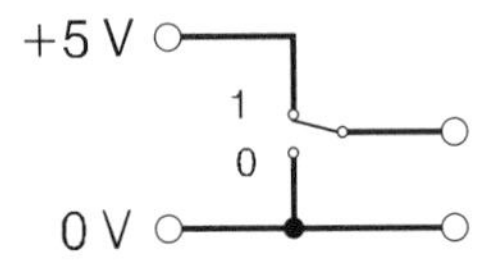

Figure 1.3 *Simple pulse generator*

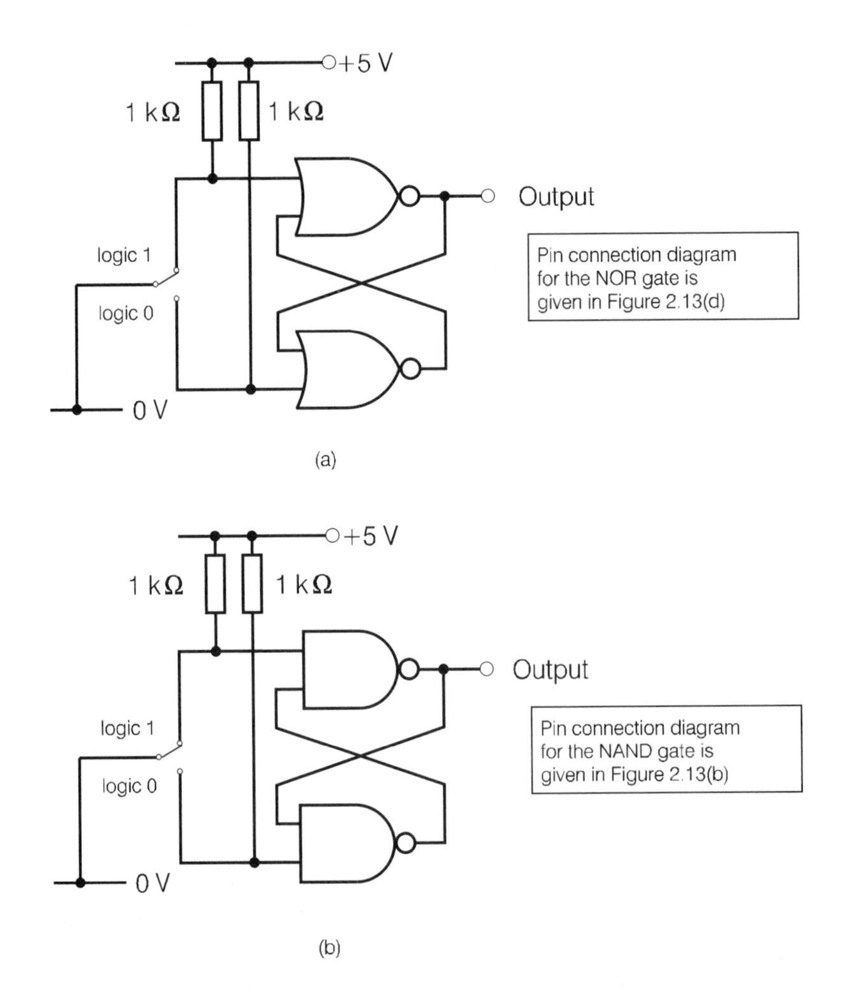

Figure 1.4 *The de-bounced switch: (a) using NOR gates (b) using NAND gates*

Unfortunately these mechanical switches suffer from an effect called **contact bounce**. When the switches are operated as an intended one-off change, the contacts can make and break several times with each make and break counting as a pulse. The result, as far as the system is concerned, is erratic and unsatisfactory operation. In order that the system receives just one clean pulse, the contact bounce is eliminated by using a de-bouncing circuit, as shown in Figure 1.4. The operation of the circuit is not important for the moment (it is dealt with in Chapter 6, and the information needed for its assembly is provided).

The importance of **waveform diagrams** in the understanding of the operation of digital systems will be stressed many times in this book. This brings in the need for a **cathode ray oscilloscope** (preferably double beam) and a **signal generator** providing a waveform of appropriate shape, frequency and voltage output. A sinusoidal signal is required for one section of the work; for the remainder, the need is for a **train of pulses**, with a base of 0 V and rising to +5 V. This is very adequately provided by a clock pulse generator whose circuit is shown in Figure 1.5. The same comments as above regarding the operation apply here, since this unit is dealt with in Chapter 8.

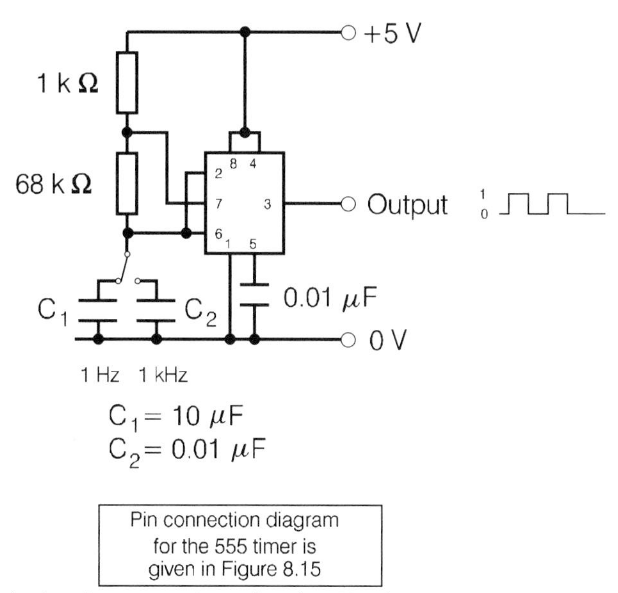

Figure 1.5 *The clock pulse generator using the 555 timer*

Circuit diagrams

The usual method here of indicating the state of the system output is by the use of LEDs. These *must* be used with a series current-limiting resistor, otherwise they will eventually be destroyed. For the particular type suggested (5 mm diameter), a resistor of value 270 Ω is suitable. Resistors are available as a dual-in-line or a single-in-line package, suitable for breadboard connection.

The LED lead connections are shown in Figure 1.6. The switches, other than those shown as providing clock pulse inputs, can be wire, connected or not according to their purpose. They also are available in dual-in-line form.

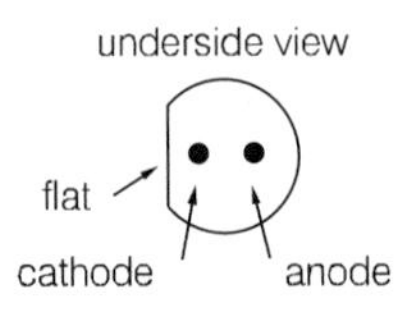

Figure 1.6 *Connections for the LED*

2 Logic gates

Logic gates are the basic building blocks of combinational logic and can have one or more inputs but **only one output**.

The algebra of logic, which is the method of relating the output to the input(s), is called Boolean algebra, after the mathematician George Boole (1815–64). This method produces an expression or statement relating the output (F, Q, Z, or really any convenient letter towards the end of the alphabet!) to the input or inputs (A, B, C etc.).

Types of logic gate

There are five basic gates, whose function is to perform logical operations. They are the AND, OR, NOT, NAND and NOR gates. Note the use of capital letters to write the name of each gate.

There are, in addition, two other gates, the XOR (exclusive OR) and the XNOR (exclusive NOR), which will be dealt with later on.

The AND gate

The principle of this gate is shown in simple series electrical circuit form in Figure 2.1. When either switch A or switch B is open, there will be no voltage (= 0 V) on the right-hand side of that particular switch, and hence **no voltage across the lamp**. There will be a voltage (= 6 V) across the lamp, causing it to light, **only** when both switch A AND switch B are closed.

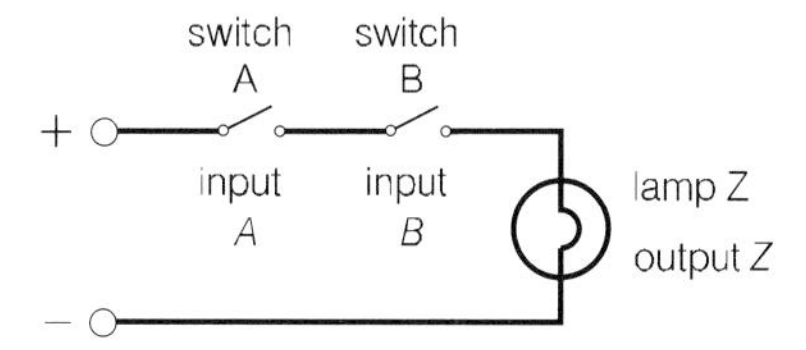

Figure 2.1 *The series electrical circuit to represent the AND gate*

An open switch (= 0 V) and a lamp which *is not lit* (= 0 V) are represented by logic level '0'.

A closed switch (= 6 V) and a lamp which *is lit* (= 6 V) are represented by logic level '1'.

For any other combination of the switch positions, the lamp will not light.

The inputs are A and B and the output is Z. The state of the output for each combination of inputs can be written in the form of a truth table, shown in Figure 2.2. From the right-hand side of the truth table it can be seen that the output Z will be logic 1 **only** when both input A AND input B are logic 1. This can be written in Boolean algebra form as a logic statement or expression, given by

$$Z = A \cdot B$$

where '·' means 'AND'.

INPUTS		OUTPUT
switch		lamp
A	B	Z
open (0)	open (0)	not lit (0)
open (0)	closed (1)	not lit (0)
closed (1)	open (0)	not lit (0)
closed (1)	closed (1)	lit (1)

Figure 2.2 *Truth table for the series electrical circuit*

The above example relates to a 2-input AND gate, the circuit symbol and truth table for which are shown in Figure 2.3.

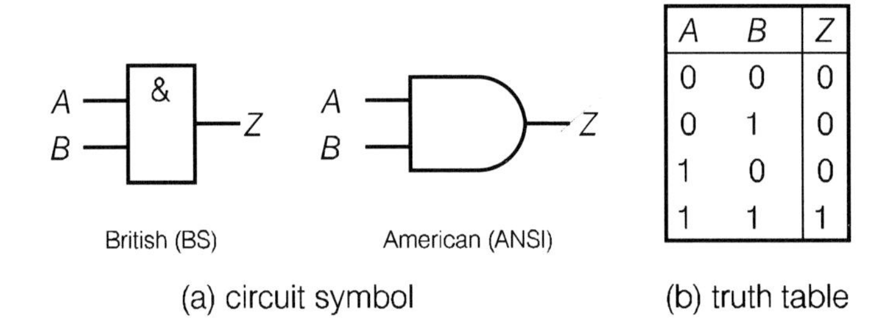

A	B	Z
0	0	0
0	1	0
1	0	0
1	1	1

(b) truth table

Figure 2.3 *The 2-input AND gate: circuit symbol and truth table*

Important points

- BS means British Standard. The British Standard for circuit symbols is BS 3939. ANSI stands for American National Standards Institute.
- In practice, the American symbols appear more frequently in magazines etc. Many people (students included!) feel that these are clearer and easier to remember.

- AND gates are available with a larger number of inputs, for example, three or four.
- From now on the phrase 'Z is logic 1' will be written 'Z is 1'.

The OR gate

A similar, but parallel rather than series, circuit can be used to represent the OR gate and is shown in Figure 2.4, with its truth table in Figure 2.5. Referring again to the right-hand side of the truth table, it can be seen now that the output Z is 1 when either input A OR input B is 1; and also, incidentally, when both inputs are 1.

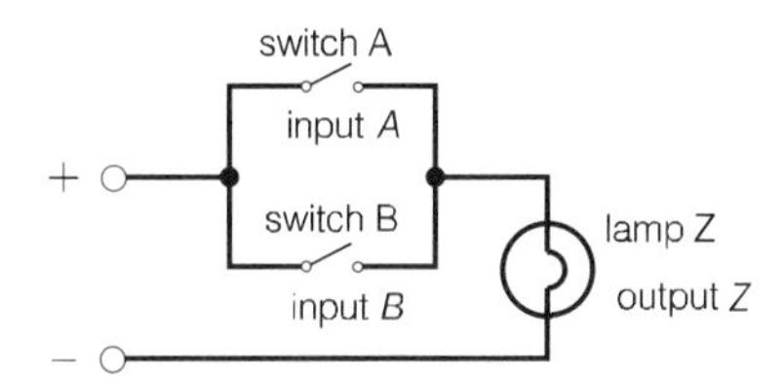

Figure 2.4 *The parallel electrical circuit to represent the OR gate*

INPUTS				OUTPUT	
switch				lamp	
A		B		Z	
open	(0)	open	(0)	not lit	(0)
open	(0)	closed	(1)	lit	(1)
closed	(1)	open	(0)	lit	(1)
closed	(1)	closed	(1)	lit	(1)

Figure 2.5 *Truth table for the parallel electrical circuit*

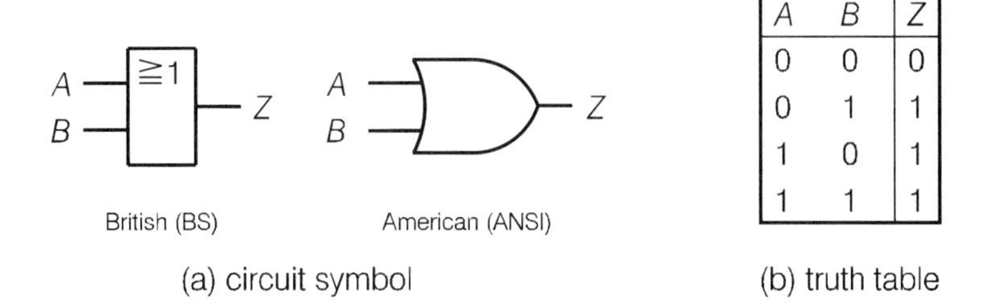

A	*B*	*Z*
0	0	0
0	1	1
1	0	1
1	1	1

Figure 2.6 *The 2-input OR gate: circuit symbol and truth table*

In Boolean algebra, the logic expression for this gate is given as

$$Z = A + B$$

where '+' means 'OR'.
The circuit symbol and truth table are shown in Figure 2.6.

The NOT gate

For those interested in the 'simple' circuit, a common emitter transistor amplifier will perform this function. The circuit symbol and truth table are shown in Figure 2.7.

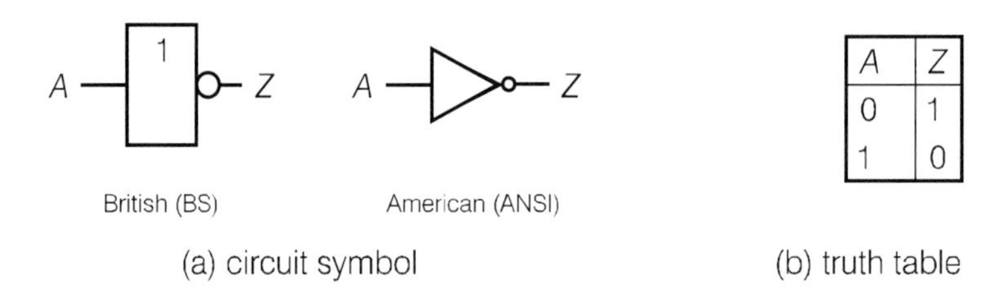

A	Z
0	1
1	0

(b) truth table

Figure 2.7 *The NOT gate (inverter): circuit symbol and truth table*

Important points

- The output of the NOT gate is always the opposite (the inverse) of the input. Hence the NOT gate is also known as an **inverter**.
- The NOT gate will always have just the **one input**.
- Notice the bubble on the right-hand side of the symbol, which denotes the NOT function.

The logic (Boolean) expression for the NOT gate is given by

$Z = \overline{A}$ (called '*A* bar')

where '–' means 'NOT'.

The NAND gate

This gate is the inverse of the AND gate, that is, the opposite as far as the output is concerned. The symbol and truth table are shown in Figure 2.8.

In this case, the logic expression is given by

$$Z = \overline{A \cdot B}$$

or, in words, '*Z* is 1 when (both) *A* AND *B* are NOT 1', which is illustrated by the first three lines in the truth table.

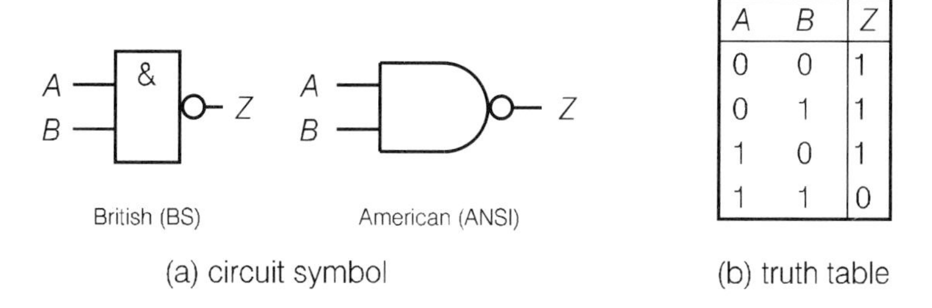

A	B	Z
0	0	1
0	1	1
1	0	1
1	1	0

Figure 2.8 *The 2-input NAND gate: circuit symbol and truth table*

An easier alternative is 'Z is NOT (A AND B)'.

The last line in the truth table tells us that 'Z is NOT 1 when A AND B are both 1'. This will give what is known as the inverse function for Z, namely $\overline{Z}$, which is given by

$$\overline{Z} = A \cdot B$$

In any case it may be easier to remember that the NAND output is the opposite of the AND!

> **Important points**
>
> - The 'bar-line' must be a complete line over $A.B$ and not separate lines. You will show for yourself later on that
>
> $\overline{A} \cdot \overline{B}$ is **not the same** as $\overline{A \cdot B}$
>
> - Again notice the bubble.

The NOR gate

This gate is the inverse of the OR gate. Figure 2.9 shows the circuit symbol and truth table.

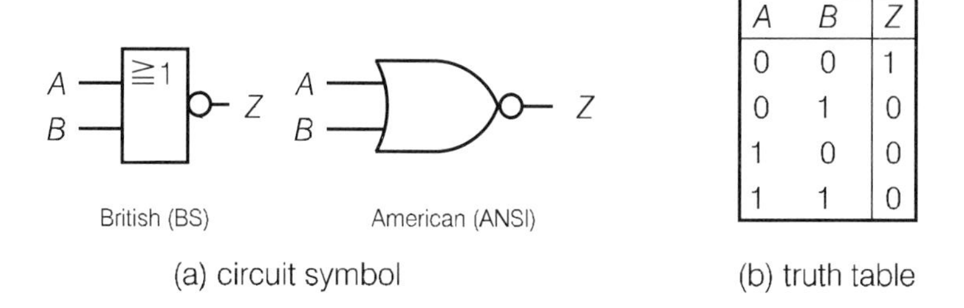

A	B	Z
0	0	1
0	1	0
1	0	0
1	1	0

Figure 2.9 *The 2-input NOR gate: circuit symbol and truth table*

The logic expression is given by

$$Z = \overline{A + B}$$

> **Important points**
>
> - As with the NAND gate, the bar-line must be complete. Again you will find that
>
> $Z = \overline{A} + \overline{B}$ is **not the same** as $Z = \overline{A + B}$
> - A reminder from the OR gate truth table (Figure 2.6) that the output Z is also 1 when both inputs A and B are 1.
>
>   There may be a need in practice for the output of the OR gate to be 0 when both inputs are 1. In other words, we may wish to exclude the situation in which all inputs at logic 1 give an output of logic 1.

Exclusive OR and exclusive NOR gates

This leads us to the exclusive OR gate (abbreviated to XOR) whose circuit symbol and truth table are given in Figure 2.10.

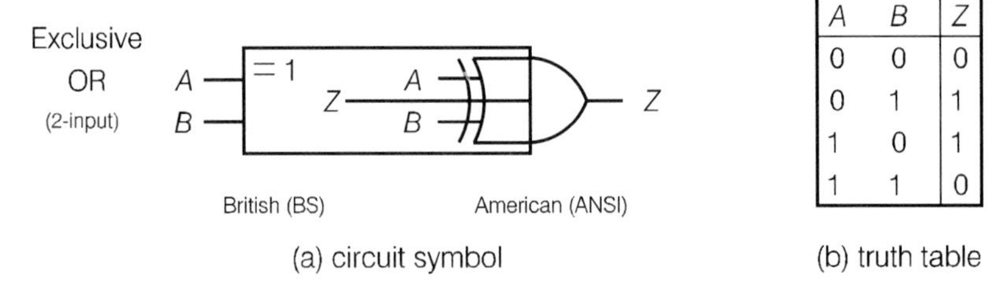

A	B	Z
0	0	0
0	1	1
1	0	1
1	1	0

(b) truth table

Figure 2.10 *The exclusive OR gate: circuit symbol and truth table*

The logic expression for the 2-input exclusive OR gate is given by

$$Z = A \oplus B$$

Alternatively, from the truth table, we see that

$$Z = 1 \text{ when } A = 0, B = 1, \text{ which gives } Z = \overline{A} \cdot B$$

and

$$Z = 1 \text{ when } A = 1, B = 0, \text{ which gives } Z = A \cdot \overline{B}$$

The expression for Z then becomes $Z = \overline{A} \cdot B + A \cdot \overline{B}$.

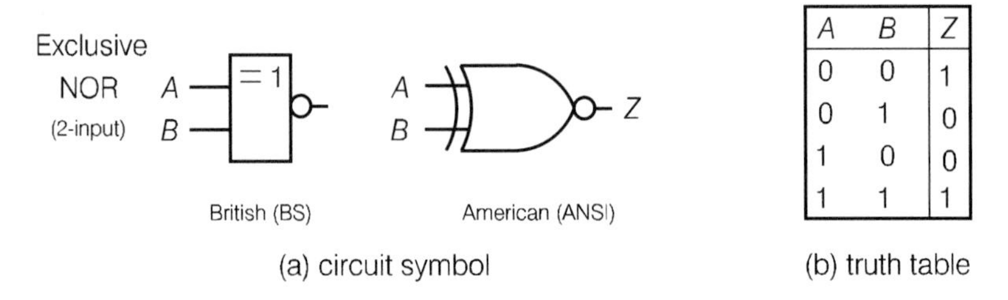

A	B	Z
0	0	1
0	1	0
1	0	0
1	1	1

(b) truth table

Figure 2.11 *The exclusive NOR gate: circuit symbol and truth table*

There is (inevitably) the inverse of the exclusive OR, called the exclusive NOR, see Figure 2.11.

A summary of these gates is given in Figure 2.12.

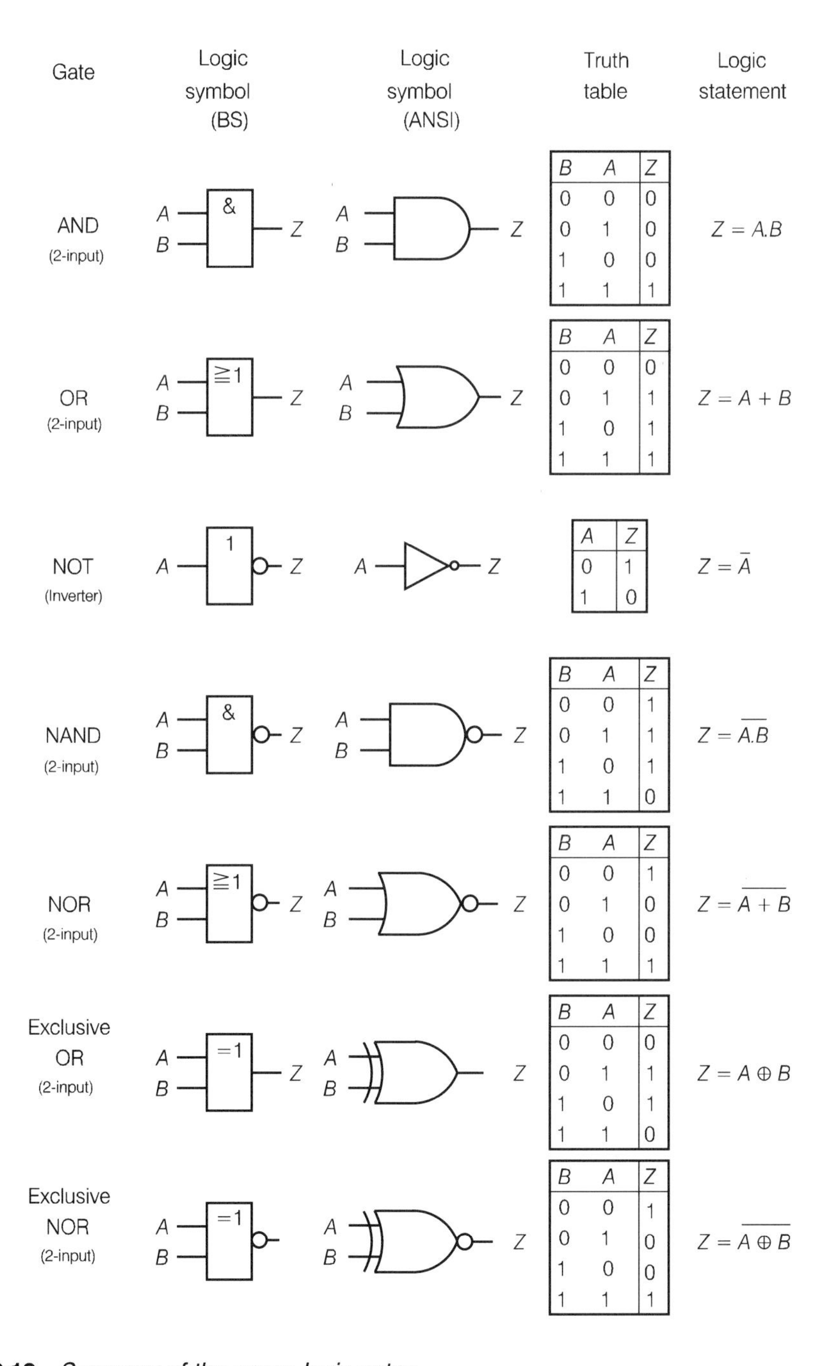

Gate	Truth table (B A Z)	Logic statement
AND (2-input)	0 0 0; 0 1 0; 1 0 0; 1 1 1	$Z = A.B$
OR (2-input)	0 0 0; 0 1 1; 1 0 1; 1 1 1	$Z = A + B$
NOT (Inverter)	A Z: 0 1; 1 0	$Z = \overline{A}$
NAND (2-input)	0 0 1; 0 1 1; 1 0 1; 1 1 0	$Z = \overline{A.B}$
NOR (2-input)	0 0 1; 0 1 0; 1 0 0; 1 1 1	$Z = \overline{A + B}$
Exclusive OR (2-input)	0 0 0; 0 1 1; 1 0 1; 1 1 0	$Z = A \oplus B$
Exclusive NOR (2-input)	0 0 1; 0 1 0; 1 0 0; 1 1 1	$Z = \overline{A \oplus B}$

Figure 2.12 *Summary of the seven logic gates*

Logic gate packages

It will be useful now if you can get hold of data sheets for these logic gates. This information is readily available in various suppliers' catalogues (e.g. Maplin and RS Components). The details given below illustrate the explanations:

Device number	Type	Number of pins
7408 (or 74LS08)	Quad 2-input AND gate	14
7421 (74LS21)	Dual 4-input AND gate	14
7411 (74LS11)	Triple 3-input AND gate	14
7404 (74LS04)	Hex inverter	14

Important points

- Logic gates are available in packages; that is, a number of gates of the same type are contained within the single chip. Thus we see that the integrated circuit (ic) package numbered 7408 contains four AND gates, each gate having two inputs. The 7408 ic chip is thus described as a

  'quad (meaning four) 2-input AND'

  Similarly, the 7421 ic is a

  'dual (meaning two) 4-input AND'

  It will be obvious that the greater the number of inputs required for the particular gate, the smaller the number of gates there can be in any one package.
- The 7404 'hex inverter' is an ic containing six (hex) NOT (inverter) gates.
- The pin arrangement (or pin-out) is known as an *n*-pin dual-in-line (dil) package, where *n* will be 8, 14, 16 etc. as necessary. Thus the 7408 ic is a '14-pin dil'.

Practical Exercise 2.1

To become familiar with the five basic logic gates
For this exercise you will need the following components and equipment:

1 – 74LS08 ic (quad 2-input AND)
1 – 74LS32 ic (quad 2-input OR)
1 – 74LS04 ic (hex inverter) (NOT)
1 – 74LS00 ic (quad 2-input NAND)
1 – 74LS02 ic (quad 2-input NOR)
1 – LED (5 mm) and series resistor (270 Ω)
1 – +5 V DC power supply
1 – multimeter

The pin connection diagrams for the ic's are given in Figure 2.13.

The general circuit arrangement, using the AND gate as the example, is shown in Figure 2.14.

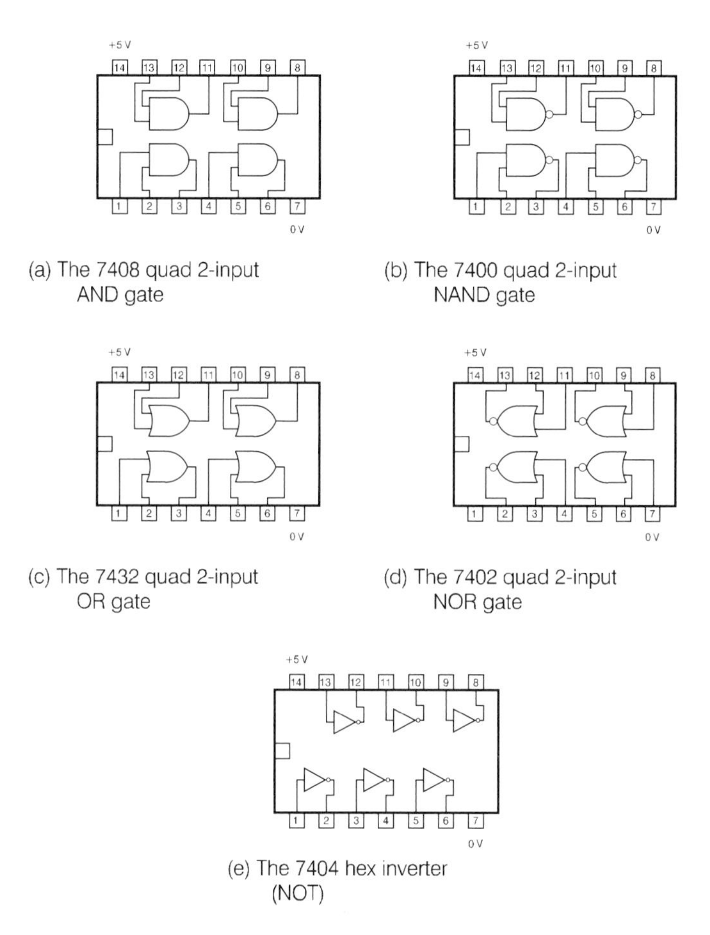

(a) The 7408 quad 2-input AND gate

(b) The 7400 quad 2-input NAND gate

(c) The 7432 quad 2-input OR gate

(d) The 7402 quad 2-input NOR gate

(e) The 7404 hex inverter (NOT)

Figure 2.13 *The five basic gates: pin connections*

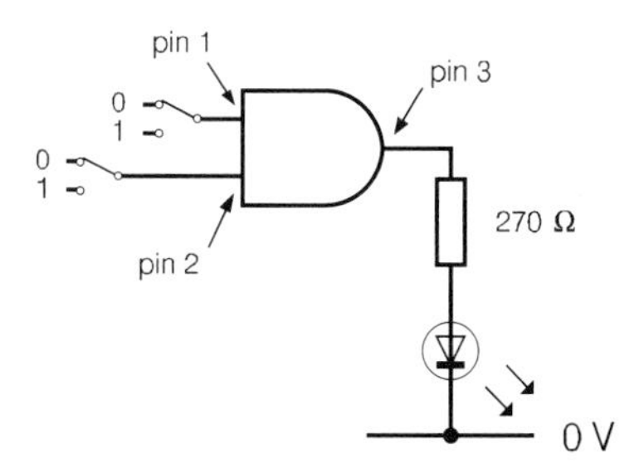

Figure 2.14 *Circuit arrangement for the AND gate: Practical Exercise 2.1*

Continued on p. 16

Practical Exercise 2.1 *(Continued)*

Procedure

1 Taking each ic in turn, choose one gate from within the package, connect it to the supply and check that it performs according to the appropriate truth table.

Make sure that all inputs are connected (to either 0 or 1) as necessary, and not left floating.

2 Measure the current taken from the supply by each chip. (To do this, just open-circuit the positive supply line to the chip and connect the ammeter in the break.)

3 To investigate the effect of leaving an input in the so-called 'floating' condition, reconnect the AND gate to the supply, leave both inputs disconnected and determine the output logic level. (You should find that the output is 1.)

Now, with the inputs still floating, determine the logic level at the input pins of the ic. To do this, you will need to measure the DC voltage at these pins.

From this you should now see why the output was 1!

Questions

2.1 Give the ANSI logic symbol, the logic expression and the truth table, for
   (a) the 3-input AND gate
   (b) the 3-input NOR gate

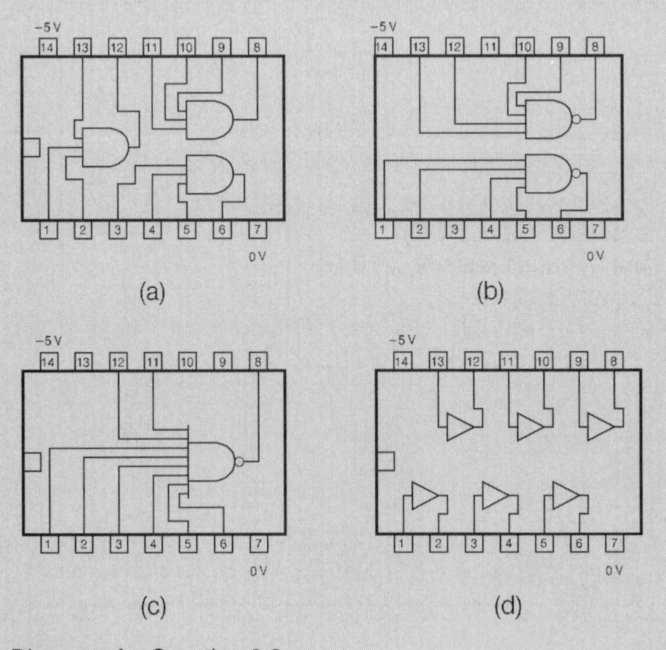

Figure 2.15 *Diagrams for Question 2.2*

2.2 Use the conventional system (stating the number of gates followed by the number of inputs for each gate) to describe each of the logic ic's shown in Figure 2.15.

2.3 Draw up the truth table for each of the logic expressions

$Z = \overline{A \cdot B}$ and $Z = \overline{A} \cdot \overline{B}$

and hence show that these two expressions *are not* identical.

2.4 Draw up the truth table for each of the logic expressions

$Z = \overline{A + B}$ and $Z = \overline{A} + \overline{B}$

and hence show that these two expressions *are not* identical.

2.5 Draw up the truth table for each of the logic expressions

$Z = \overline{A} \cdot B + A \cdot \overline{B} + A \cdot B$ and $Z = A + B$

and hence show that these two expressions *are* identical.

2.6 What logic level does an unconnected input normally assume?

So far we have seen that the five basic gates are all available in separate packages, e.g. 7408 = AND, 7400 = NAND and so on.

As a result of doing Practical Exercise 2.2 which follows, it will be found that a combination of NAND (or of NOR) gates can be used to provide any of the other logic functions.

Practical Exercise 2.2

To investigate the use of the universal NAND and NOR gates
For this exercise you will need the following components and equipment:

1 – 74LS00 ic (quad 2-input NAND)
1 – 74LS02 ic (quad 2-input NOR)
1 – LED (5 mm) + series resistor (270 Ω)
1 – +5 V DC power supply

The pin connection diagrams are given in Figure 2.13.

Procedure

1 See Figure 2.16. Connect up each circuit separately, using a similar arrangement to that shown in Figure 2.14. As a result, produce the truth table and the logic expression for each.

Continued on p. 18

Practical Exercise 2.2 (*Continued*)

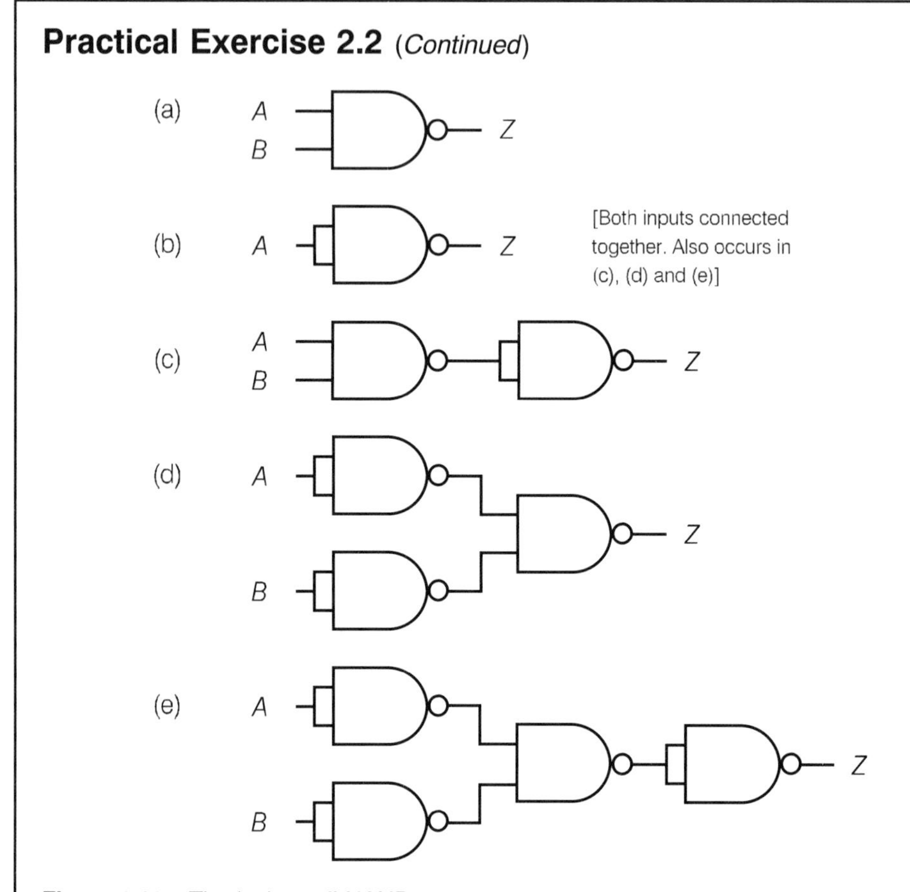

Figure 2.16 *The 'universal' NAND gate*

You should find that each circuit will provide one different logic function and that all five functions (AND, OR, NOT, NAND and NOR) can be made up using just one type of logic gate (NAND).

2 Repeat procedure 1, but now with NOR instead of NAND gates, using the same circuit arrangement (Figures 2.16 and 2.14).

You should again find that all five gates can be made up.

N.B. The provision of these five functions may be in a different order from those for the NAND gates!

Question

2.7 What are the advantages in constructing a logic circuit entirely of either NAND or NOR gates?

Multiple input logic gates

An inspection of suppliers' catalogues will reveal the availability of gates having more than two inputs. These multiple input gates will thus be very convenient to use where space is at a premium. When a particular, otherwise non-available gate is required, it is a fairly straightforward task to provide this, using a combination of, say, 2-input gates. The following examples should help to make this clear. The reader is asked to note the use of the truth table to verify each requirement.

Example (a)

To make up a 3-input AND gate using 2-input AND gates.

The required logic diagram and related truth table are shown in Figure 2.17.

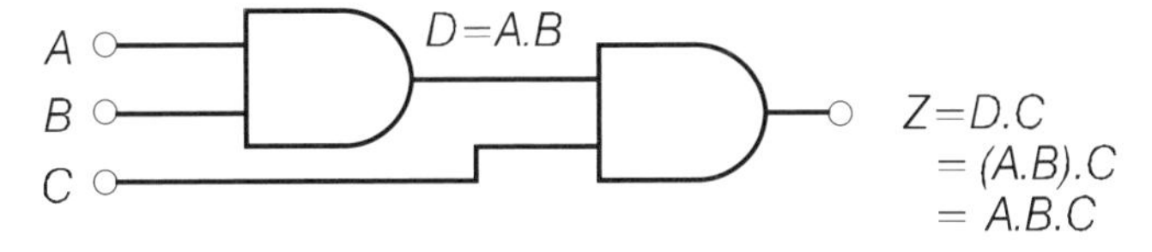

(a) Logic diagram

A	B	C	D=A.B	Z=D.C =(A.B).C =A.B.C
0	0	0	0	0
0	0	1	0	0
0	1	0	0	0
0	1	1	0	0
1	0	0	0	0
1	0	1	0	0
1	1	0	1	0
1	1	1	1	1

(b) Truth table

Figure 2.17 *A 3-input AND gate using 2-input AND gates: logic diagram and truth table*

Example (b)

To make up a 4-input AND gate using 2-input AND gates.

The required logic diagram is shown in Figure 2.18.

The reader is invited to draw up the truth table in order to verify this arrangement.

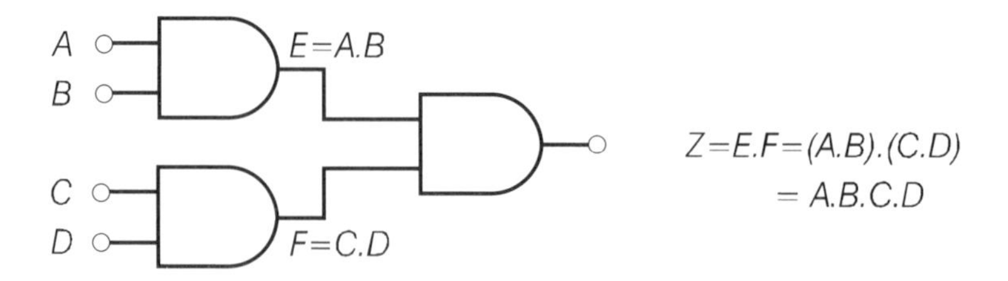

Figure 2.18 *A 4-input AND gate using 2-input AND gates: logic diagram*

Example (c)

To make up a 3-input OR gate using 2-input OR gates.
The required logic diagram is shown in Figure 2.19.

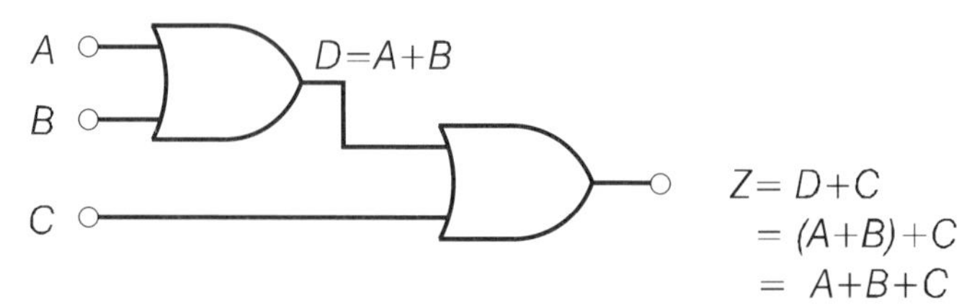

Figure 2.19 *A 3-input OR gate using 2-input OR gates: logic diagram*

Questions

Draw a logic diagram to show how each of the following multiple input gates can be made up using the appropriate 2-input gates. Verify your arrangements with truth tables.

2.8 A 5-input AND gate
2.9 A 4-input OR gate
2.10 A 5-input OR gate

Example (d)

To (attempt) to make up a 3-input NAND gate using 2-input NAND gates.

Let us try the method used earlier. Figure 2.20(a) shows a suggested logic diagram. The truth table of Figure 2.20(b) shows that this arrangement is unsuccessful.

The correct arrangement, shown in Figure 2.21(a), requires a 2-input OR gate **in addition to** the two 2-input NAND gates. The reader will notice the use of one NAND gate as a NOT gate. The truth table given in Figure 2.21(b) shows the justification.

A similar problem will occur in trying to make up a 3-input NOR gate from two 2-input NOR gates. A similar solution will require the addition of a 2-input AND gate.

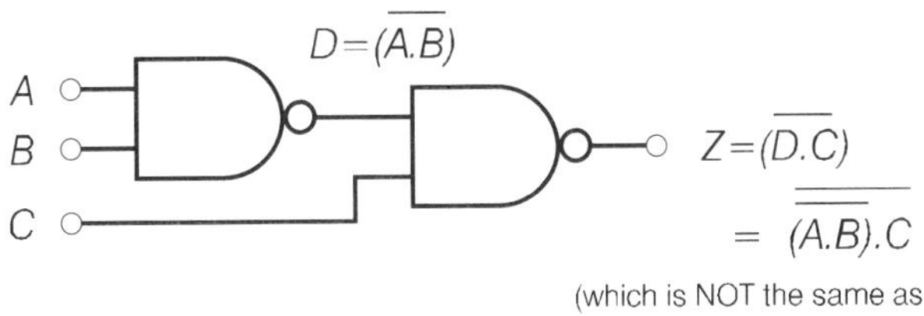

(which is NOT the same as $\overline{A.B.C}$ – see truth table)

(a) Logic diagram

A	B	C	$A.B$	$D=\overline{A.B}$	$D.C$	$Z=\overline{D.C}$ $=\overline{(\overline{A.B}).C}$	$\overline{A.B.C}$
0	0	0	0	1	0	1	1
0	0	1	0	1	1	0	1
0	1	0	0	1	0	1	1
0	1	1	0	1	1	0	1
1	0	0	0	1	0	1	1
1	0	1	0	1	1	0	1
1	1	0	1	0	0	1	1
1	1	1	1	0	0	1	0

(b) Truth table

Figure 2.20 *An attempt to make a 3-input NAND gate: logic diagram and truth table*

Questions

2.11 You have 2-input NOR gates and 2-input AND gates available. Draw the logic diagram for a 3-input NOR gate and justify your answer with a truth table.

2.12 Draw the logic diagram for
- (a) a 4-input NAND gate
- (b) a 4-input NOR gate.

Logic rules and theorems

You will discover as you follow through the table in Figure 2.22 that the majority of the so-called rules are in fact fairly obvious, so that you may wonder why they need to be set out.

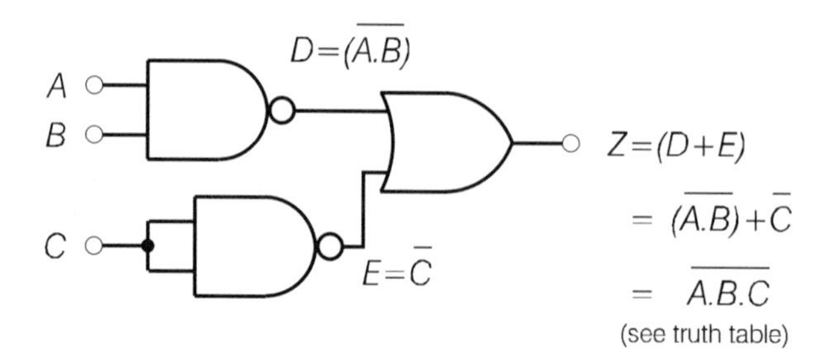

(a) Logic diagram

A	B	C	A.B	$D=\overline{A.B}$	$\overline{C}$	$Z=D+\overline{C}$ $=\overline{(A.B)}+\overline{C}$	$\overline{A.B.C}$
0	0	0	0	1	1	1	1
0	0	1	0	1	0	1	1
0	1	0	0	1	1	1	1
0	1	1	0	1	0	1	1
1	0	0	0	1	1	1	1
1	0	1	0	1	0	1	1
1	1	0	1	0	1	1	1
1	1	1	1	0	0	0	0

(b) Truth table

Figure 2.21 *A 3-input NAND gate: logic diagram and truth table*

They can all be proved in at least two ways:

- by practical exercises
- by drawing up a truth table.

Before embarking on the next practical exercise let us have a look at a couple of these rules/theorems and see how they can be justified, by drawing up the truth table.

Firstly, let us dispose of Rules 1 to 4 by suggesting that they are merely saying that it does not matter in which order you perform the function. In other words, Rule 1 is saying

'*A* OR *B* is **the same as** *B* OR *A*'

and so on.

Secondly, Rule 5 uses ordinary algebra, in which every term inside the bracket, i.e. *B* and *C*, is multiplied by the term(s) outside the bracket, in this case, just *A*.

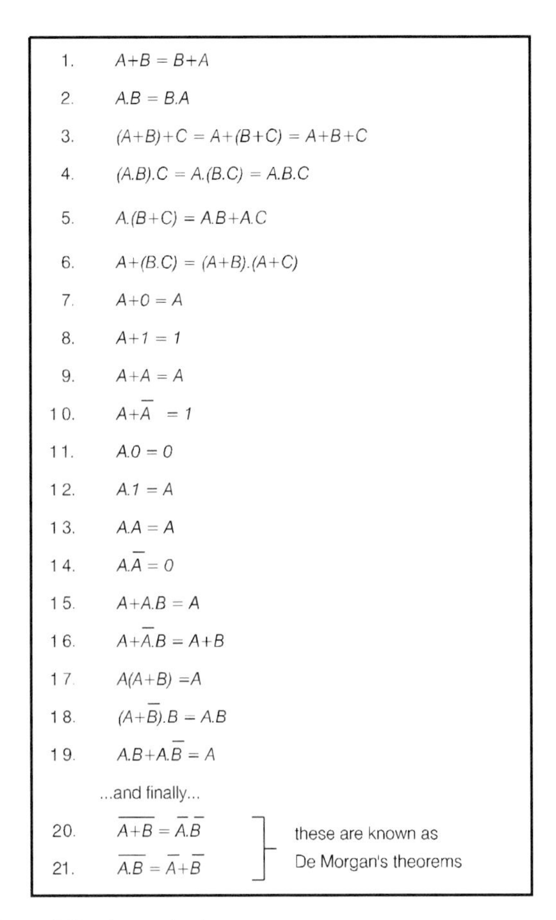

1. $A+B = B+A$
2. $A.B = B.A$
3. $(A+B)+C = A+(B+C) = A+B+C$
4. $(A.B).C = A.(B.C) = A.B.C$
5. $A.(B+C) = A.B+A.C$
6. $A+(B.C) = (A+B).(A+C)$
7. $A+0 = A$
8. $A+1 = 1$
9. $A+A = A$
10. $A+\overline{A} = 1$
11. $A.0 = 0$
12. $A.1 = A$
13. $A.A = A$
14. $A.\overline{A} = 0$
15. $A+A.B = A$
16. $A+\overline{A}.B = A+B$
17. $A(A+B) = A$
18. $(A+\overline{B}).B = A.B$
19. $A.B+A.\overline{B} = A$

...and finally...

20. $\overline{A+B} = \overline{A}.\overline{B}$
21. $\overline{A.B} = \overline{A}+\overline{B}$

(20 and 21: these are known as De Morgan's theorems)

Figure 2.22 *The rules of Boolean algebra*

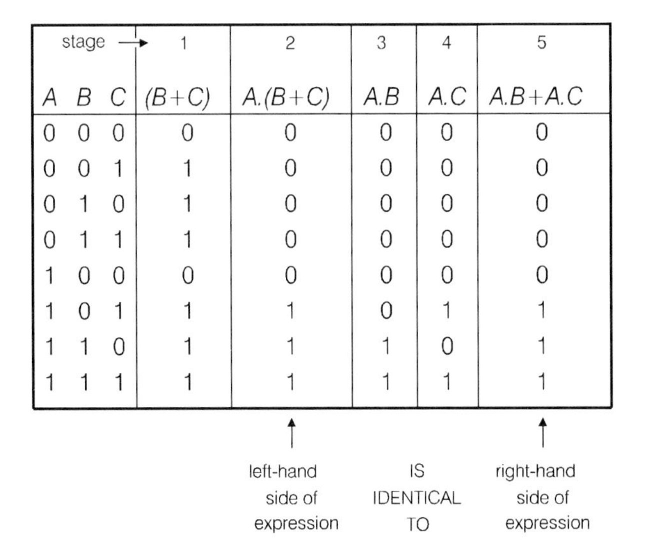

stage →			1	2	3	4	5
A	B	C	(B+C)	A.(B+C)	A.B	A.C	A.B+A.C
0	0	0	0	0	0	0	0
0	0	1	1	0	0	0	0
0	1	0	1	0	0	0	0
0	1	1	1	0	0	0	0
1	0	0	0	0	0	0	0
1	0	1	1	1	0	1	1
1	1	0	1	1	1	0	1
1	1	1	1	1	1	1	1

Figure 2.23 *Truth table to verify* $A \cdot (B + C) = A \cdot B + A \cdot C$

Staying with Rule 5: the truth table in Figure 2.23 has been built up in stages before arriving at the 'final' column. Doing it this way means that far less has to be remembered, thus giving less opportunity for errors.

Practical Exercise 2.3

To verify De Morgan's theorems
For this exercise you will need the following components and equipment:

1 – 74LS08 ic (quad 2-input AND)
1 – 74LS04 ic (hex inverter) (NOT)
1 – 74LS32 ic (quad 2-input OR)
1 – LED (5 mm) and series resistor (270 Ω)
1 – +5 V DC power supply

Procedure

The pin connection diagrams are given in Figure 2.13.

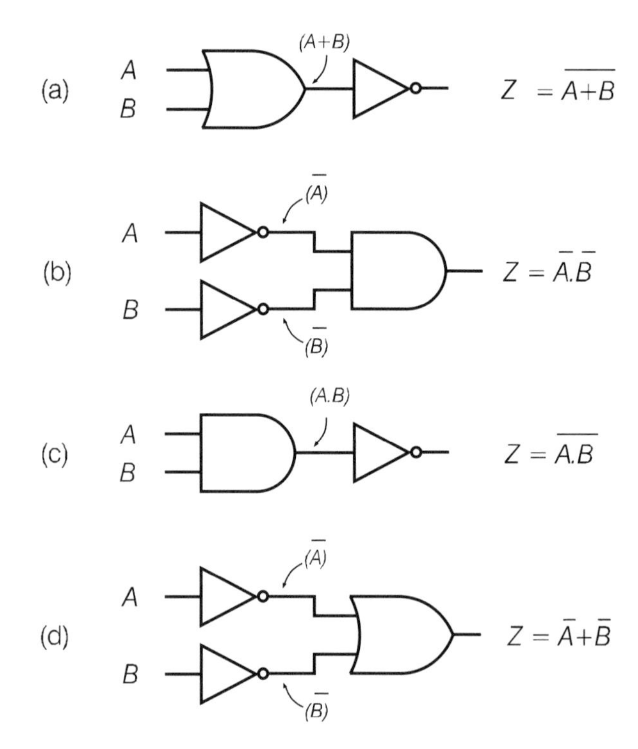

Figure 2.24 *Circuits to demonstrate De Morgan's theorems: Practical Exercise 2.3*

1 To verify $\overline{A+B} = \overline{A} \cdot \overline{B}$, connect up circuits (a) and (b) in Figure 2.24.
2 Go through the sequence for inputs A and B, and show that De Morgan's theorem holds good.
3 To verify $\overline{A \cdot B} = \overline{A} + \overline{B}$, connect up circuits (c) and (d) in Figure 2.24.
4 Go through the sequence for inputs A and B, and again show that De Morgan's theorem holds good.

Questions

2.13 Show how the exclusive OR function can be provided by the use of
(a) AND, OR and NOT gates
(b) NAND gates only
(Take the XOR expression as $Z = \overline{A} \cdot B + A \cdot \overline{B}$)

2.14 Draw up a truth table to show that
(a) $A \cdot \overline{B} + \overline{A} \cdot B = (A + B) \cdot (\overline{A} + \overline{B})$
(b) $A \cdot B + \overline{A} \cdot \overline{B} = (A + \overline{B}) \cdot (\overline{A} + B)$

Important point

- In using NAND (or NOR) gates to replace a system made from AND/OR/NOT functions you will usually find an example of **redundant** gates. This occurs when two identical NAND (or NOR) gates, being used as NOT gates, are in series, producing a double negative. The system is simplified by removing these two gates.

Simplification of logic expressions

The rules of Boolean algebra (Figure 2.22) can be used as a first attempt in the simplification process as shown in the following examples:

Example (e) $Z = (A + \overline{B}) \cdot (\overline{A} + B)$

We can use the rules of **ordinary** algebra to multiply out the brackets.

Thus

$$Z = A \cdot \overline{A} + \overline{A} \cdot \overline{B} + A \cdot B + B \cdot \overline{B}$$

$$= 0 + \overline{A} \cdot \overline{B} + A \cdot B + 0 \qquad \text{(rule 14)}$$

Giving

$$Z = \overline{A} \cdot \overline{B} + A \cdot B$$

which is the required simplification.

It will be no surprise to learn that the simplification can be proved by drawing up the appropriate truth table, as shown in Figure 2.25:

A	B	A	B	$A\bar{A}$	$B\bar{B}$	AB	$\bar{A}\bar{B}$	$(\bar{A}\bar{B}+AB)$	$(A+\bar{B})$	$(\bar{A}+B)$	$(A+\bar{B}).(\bar{A}+B)$
0	0	1	1	0	0	0	1	1	1	1	1
0	1	1	0	0	0	0	0	0	0	1	0
1	0	0	1	0	0	0	0	0	1	0	0
1	1	0	0	0	0	1	0	1	1	1	1

these are identical

Figure 2.25 *Truth table for example (e)*

Example (f) $Z = A \cdot B + B + C$

Using ordinary algebra, this time to take common terms **outside** a bracket, we get

$$Z = B(A+1) + C$$

$$A + 1 = 1 \qquad \text{(rule 8)}$$

Thus $\quad Z = B + C$

which is the required simplification.

It will often be helpful, and indeed necessary, to use De Morgan's theorems during the simplification process, as the following examples (g) and (h) will show:

Example (g)

Simplify $Z = (\overline{A \cdot B}) \cdot (A \cdot B + B \cdot C)$

The use of De Morgan's theorem, $\overline{A \cdot B} = \bar{A} + \bar{B}$ (rule 21) gives

$$Z = (\bar{A} + \bar{B}) \cdot (A \cdot B + B \cdot C)$$

Multiplying out the brackets gives

$$Z = \bar{A} \cdot A \cdot B + \bar{A} \cdot B \cdot C + \bar{B} \cdot A \cdot B + \bar{B} \cdot B \cdot C$$

From rule 14,

$$\bar{A} \cdot A \cdot B = 0 \cdot B = 0$$

$$\bar{B} \cdot A \cdot B = 0 \cdot A = 0$$

$$\bar{B} \cdot B \cdot C = 0 \cdot C = 0$$

Thus

$$Z = \bar{A} \cdot B \cdot C$$

which is the required simplification.

Example (h)

Show that the following expressions are identical:

$$Z = (\overline{A \cdot B}) + \overline{C} \text{ and } Z = \overline{A \cdot B \cdot C}$$

This is the 3-input NAND gate of Example (d)! Let us try using De Morgan's theorem(s) as follows:

$$\begin{aligned} Z &= (\overline{A \cdot B}) + \overline{C} \\ &= (\overline{A} + \overline{B}) + \overline{C} \quad \text{(rule 21)} \\ &= (\overline{A} + \overline{B} + \overline{C}) \quad \text{(rule 3)} \\ &= (\overline{A \cdot B \cdot C}) \quad \text{(rule 21)} \end{aligned}$$

which is the required solution.

Questions

Use the rules of Boolean algebra to simplify the following expressions and verify each simplification by drawing up a truth table:

2.15 $Z = A \cdot (\overline{A} + B)$
2.16 $Z = A + A \cdot B + A \cdot C$
2.17 $Z = \overline{A} \cdot B + A \cdot \overline{B} + A \cdot B$
2.18 $Z = (A + B) \cdot (B + C)$
2.19 $Z = (\overline{A + B}) \cdot \overline{C}$
2.20 $Z = \overline{(\overline{A + B}) + B}$

3 Combinational logic

First of all it is necessary to define the term 'combinational logic'. Combinational logic is a system whereby the output condition of the system depends only on the input conditions existing at that time. This output will always be the same for that particular combination of inputs.

Important points

- Combinational logic is a decision-making system, providing a solution to a particular set of circumstances or problem.
- Combinational logic is not merely gates in combination (although they are just that) since, as we shall see later, the other system (called sequential logic) also consists of combinations of gates.

An example is shown by the logic diagram in Figure 3.1(a). In this example the state of the output (Z) is the result of the state of the combined inputs A and B.

The result can be seen either from the logic diagram or the truth table, Figure 3.1(b).

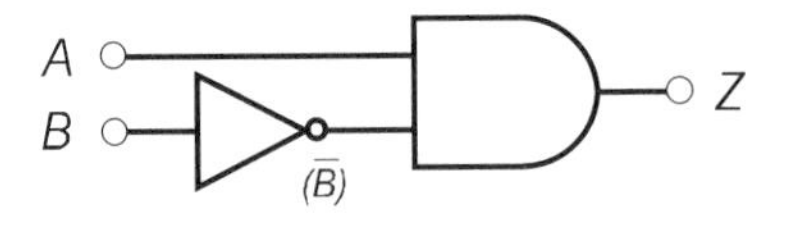

(a) Logic diagram

A	B	$\bar{B}$	Z
0	0	1	0
0	1	0	0
1	0	1	1
1	1	0	0

$Z = A.\bar{B}$

(b) Truth table

Figure 3.1 *A simple combinational logic circuit*

Important points

- It is helpful to write down the output of the NOT gate on the logic diagram, this output being one of the inputs to the AND gate.
- It is also helpful to build up the truth table in stages – in this case it only means adding the line for $\overline{B}$; hardly worth it, you may think, but it is a good habit to adopt, since all truth tables are not as simple and straightforward as this one!

Problem solving

The purpose of combinational logic is to **solve a logic problem**.

Example (a)

Consider a basic lift system. For the lift (L) to move, the following conditions have to be met:

- The lift door (D) must be closed, giving $D = 1$.
- The appropriate floor button (B) must be pressed, giving $B = 1$.

In other words, before the lift will move ($L = 1$), the door must be closed AND the button must be pressed.

The logic expression will therefore be

$$L = D \cdot B$$

with the 'problem' being solved by the use of a single AND gate.

Important points

- It is necessary at the outset of these problems, to establish a convention, which (no pun intended) will normally be 'logical'. The above example suggested that logic 1 should represent the moving lift, the closed door, and the pressed button.
  It is usual for the logic 1 condition to be given to the 'on', 'closed', 'start' etc. states.

Example (b)

A boiler shutdown solenoid (S) will operate if the temperature (T) reaches 50°C and the circulating pump (P) is turned off, or if the pilot light (L) goes out.

Setting up the conventions, let us suggest that:

- For the solenoid to operate, $S = 1$
- When the temperature reaches 50°C, $T = 1$

- When the pump is off, P = 0
- When the pilot light is out, L = 0

N.B. Logic conventions normally deal in logic 1's and so the last two states, $P = 0$ and $L = 0$, are better represented by NOT $P = 1$ and NOT $L = 1$ (i.e. by inverting both sides).

The logic expression (S) is given by

$$S = T \cdot \overline{P} + \overline{L}$$

The resulting logic diagram and truth table are shown in Figure 3.2.

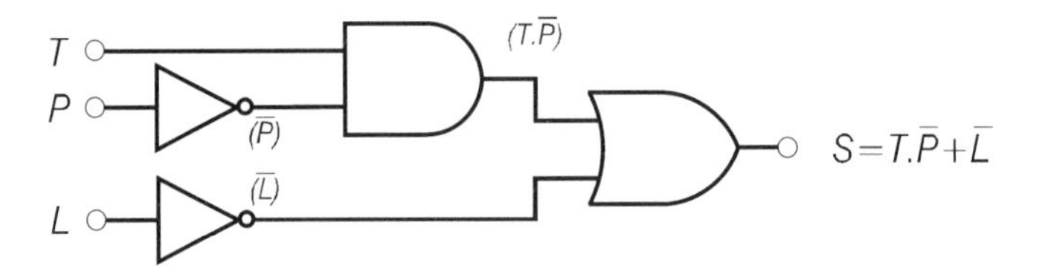

(a) Logic diagram

L	P	T	$\overline{P}$	$\overline{L}$	$T.\overline{P}$	$S=T.\overline{P}+\overline{L}$
0	0	0	1	1	0	1
0	0	1	1	1	1	1
0	1	0	0	1	0	1
0	1	1	0	1	0	1
1	0	0	1	0	0	0
1	0	1	1	0	1	1
1	1	0	0	0	0	0
1	1	1	0	0	0	0

(b) Truth table

Figure 3.2 *Logic diagram and truth table for example (b)*

Important points

- Notice again how the output side of the truth table is built up in stages, thus making it easier to obtain the correct answer for the output S.
- Care must be taken in reading (and indeed, writing) the problem in words. The placing of the 'comma' in the statement is important, since a totally different meaning would follow if the problem were written:

'... the solenoid will operate if the temperature reaches 50°C, and the pump is turned off or the pilot light goes out'.
Try it. You should arrive at a different statement for S, which now will be given by

$$S = T \cdot (\overline{P} + \overline{L})$$

(Use the truth table method to prove the difference in these statements for S!)

Questions

For each of the following problems:

- produce the logic expression
- draw up the truth table
- draw the logic diagram (called 'implementing the expression') by using:
  (a) AND, OR and NOT gates only, as necessary
  (b) NAND gates only

3.1 A machine (M) will only operate when the power (P) is on and when the safety guard (G) is in position.
3.2 A lift (L) will start if either or both of two switches X and Y are on, unless an alarm button (A) is pressed.
3.3 A system is required which will allow a door to be opened only when the correct combination of four push-buttons is pressed. Any incorrect combination will cause an alarm to ring. The four buttons are A, B, C and D and the correct combination to open the door is $A = 1$, $B = 1$, $C = 0$ and $D = 0$.
3.4 A machine operator controls red (R) and green (G) indicator lights by four switches A, B, C and D. The operating sequence is:
  (a) Red light is ON when switch A is ON and, when switch B is OFF or switch C is ON
  (b) Green light is ON when switches A and B are ON and switch C or D is OFF
3.5 A person is entitled to apply for a certain insurance policy if that person is:
  (a) a woman over the age of 40
  (b) a married man

Three switches are provided to test the entitlement:

- Switch A gives a logical 1 on the output if the person is over 40
- Switch B gives a logical 1 on the output if the person is male
- Switch C gives a logical 1 on the output if the person is married

Practical Exercise 3.1

To verify some examples of combinational logic
For this exercise you will need the following components and equipment:

2 – 74LS00 ic (quad 2-input NAND gate)
1 – 74LS10 ic (triple 3-input NAND gate)
1 – LED (5 mm) and series resistor (270 Ω)
1 – +5 V DC power supply

Procedure

1 Work through questions 1 to 5 above and justify the answers for each.
The circuit arrangement for the LED should be as Figure 2.14 but using NAND gates. The pin connections for the 74LS00 ic are given in Figure 2.13(b) and those for the 74LS10 ic in Figure 3.3.

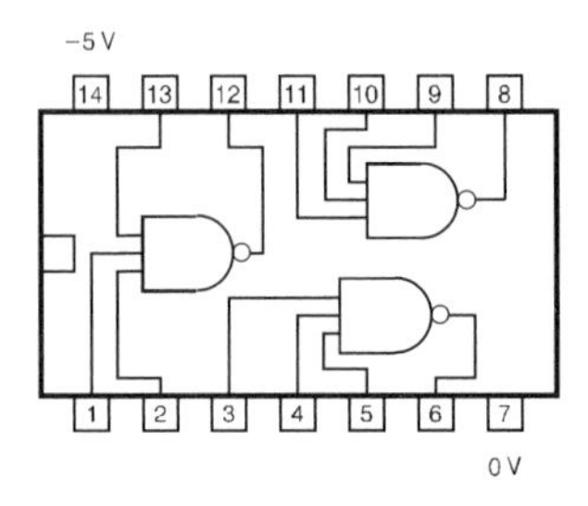

Figure 3.3 *The 7410 triple 3-input NAND gate: pin connections*

Derivation of the logic statement from the truth table

Look at the truth table in Figure 3.4. You will see that the output Z is 1 on three occasions:

(a) When input A is 0 and input B is 1: this is written as $Z = \overline{A} \cdot B$
(b) When input A is 1 and input B is 0: this is written as $Z = A \cdot \overline{B}$
(c) When input A is 1 and input B is 1: this is written as $Z = A \cdot B$

And now, since the output Z is 1 on either of these separate occasions (an OR is thus implied), the complete statement for Z can be written by connecting (a), (b) and (c) with the OR function as follows:

$$Z = \overline{A} \cdot B + A \cdot \overline{B} + A \cdot B$$

You will doubtless have recognized the truth table in Figure 3.4 as that of an OR gate. How is it, then, that the statement produced above for Z does not look like that for

A	B	Z	
0	0	0	
0	1	1	(a) $Z = \overline{A}.B$
1	0	1	(b) $Z = A.\overline{B}$
1	1	1	(c) $Z = A.B$

Figure 3.4 *Derivation of the logic statement from a truth table*

the OR gate, namely

$$Z = A + B?$$

The answer is that when a statement is derived from the truth table, a more complicated statement than necessary may be produced. But don't be alarmed! Let us have another look at the expression

$$Z = \overline{A} \cdot B + A \cdot \overline{B} + A \cdot B$$

The normal rules of algebra apply with logic functions and so we can factorize the expression for Z as follows:

$$Z = \overline{A} \cdot B + A \cdot (\overline{B} + B)$$

(factorizing the second and third terms).

From logic rule 10 (Chapter 2), we see that $(\overline{B} + B) = 1$, thus we can write

$$Z = \overline{A} \cdot B + A \cdot 1$$

From rule 12, $A \cdot 1 = A$, giving

$$Z = \overline{A} \cdot B + A$$

Finally, from rule 16, we see that $A + \overline{A} \cdot B = A + B$, thus providing the expected expression for the OR gate:

$$Z = A + B$$

All this will be regarded as laborious and it has only been pursued in order to make the point that two different-looking expressions may, in the end, be identical. 'Simplification of logic expressions' is the name of this game, but more of that in Chapter 4.

Question

3.6 For each of the following truth tables in Figure 3.5:

(a) derive the full unsimplified logic statement

(b) draw the logic diagram, using AND, OR and NOT gates only, as necessary

Continued on p. 34

Question (*Continued*)

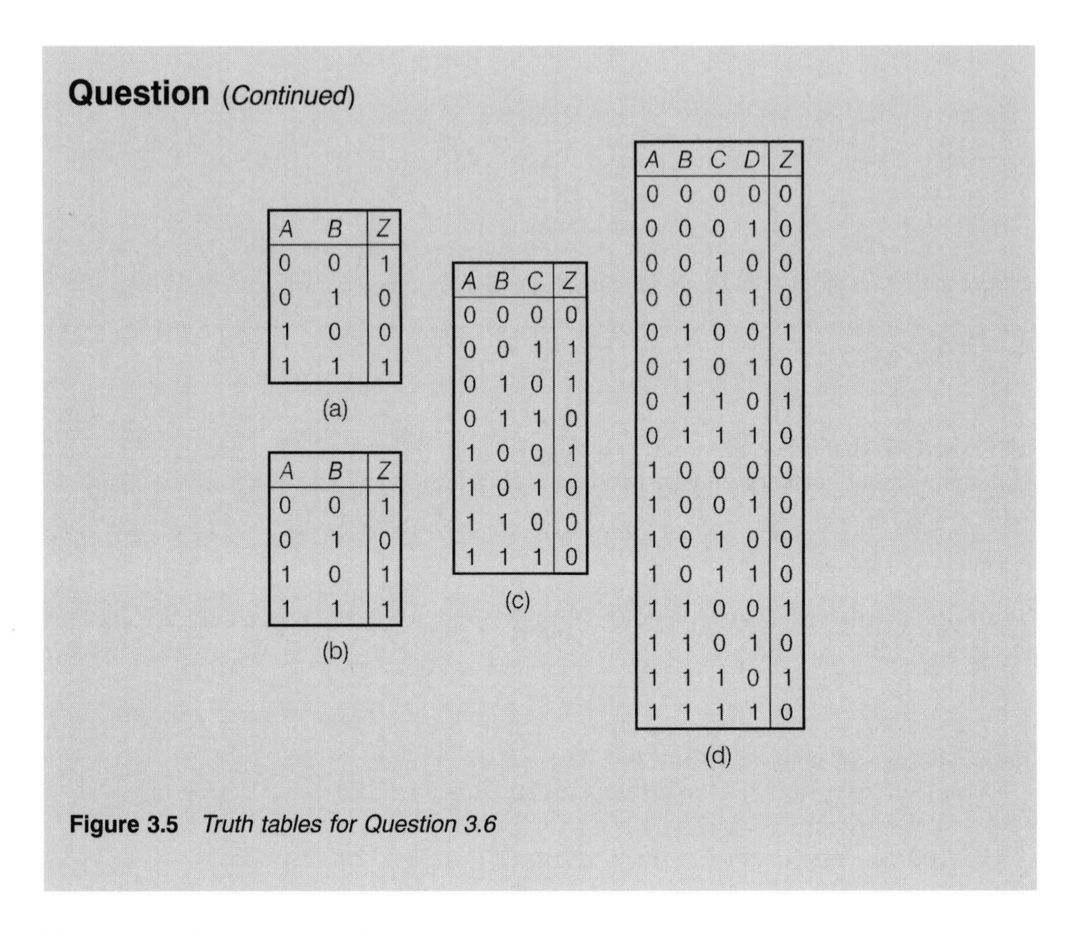

(a)

A	B	Z
0	0	1
0	1	0
1	0	0
1	1	1

(b)

A	B	Z
0	0	1
0	1	0
1	0	1
1	1	1

(c)

A	B	C	Z
0	0	0	0
0	0	1	1
0	1	0	1
0	1	1	0
1	0	0	1
1	0	1	0
1	1	0	0
1	1	1	0

(d)

A	B	C	D	Z
0	0	0	0	0
0	0	0	1	0
0	0	1	0	0
0	0	1	1	0
0	1	0	0	1
0	1	0	1	0
0	1	1	0	1
0	1	1	1	0
1	0	0	0	0
1	0	0	1	0
1	0	1	0	0
1	0	1	1	0
1	1	0	0	1
1	1	0	1	0
1	1	1	0	1
1	1	1	1	0

Figure 3.5 *Truth tables for Question 3.6*

The use of pulses as input signals

A very common application of combinational logic gates is with waveforms (known as a **train of pulses**). The next example, using Figure 3.6, will attempt to make this clear.

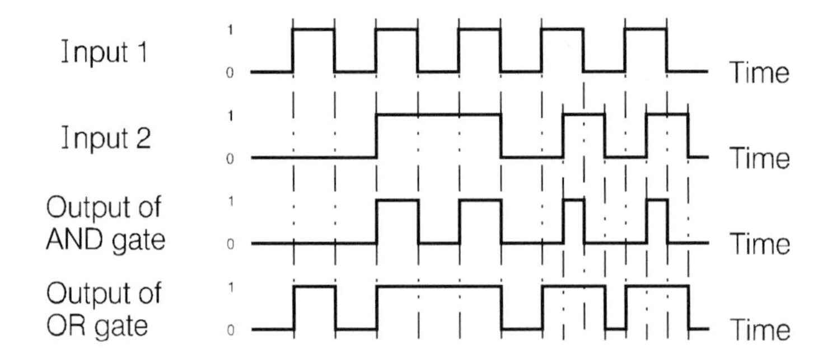

Figure 3.6 *Examples of pulse waveforms as inputs to logic gates*

Example (c)

The two input waveforms shown in Figure 3.6 are applied to

(a) a 2-input AND gate
(b) a 2-input OR gate

Draw the output waveform of each gate.

Important points

- The rules for logic gates apply. That is, for the AND gate in this example the output will be 1 *only* when both inputs are 1.

Practical Exercise 3.2

To investigate logic circuits with the use of pulse input waveforms
For this exercise you will need the following components and equipment:

1 – 74LS08 ic (quad 2-input AND)
1 – 74LS32 ic (quad 2-input OR)
1 – pulse generator (+5 V output at 1 kHz) (Figure 1.5)
1 – JK flipflop unit (74LS76 ic)
1 – double-beam cathode ray oscilloscope

Procedure

The JK flipflop unit will provide at its output a waveform whose frequency is one-half that at its input. It is known, as we shall see later on, as a **divide-by-two** device.

1 Connect up the circuit given in Figure 3.7(a) and use the double-beam oscilloscope to check that the JK flipflop is working as stated above. Draw the input and output waveforms, which should be as shown in Figure 3.7(b). Leave room for the waveform to be obtained from procedure 2.

2 Connect up the circuit shown in Figure 3.7(c) and display the output waveform together with the original (1 kHz) waveform. Draw this output waveform below those from (1) above, which you should be able to confirm as the expected output from the AND gate.

3 Repeat procedures (1) and (2), but now with the OR gate.

Continued on p. 36

Practical Exercise 3.2 (*Continued*)

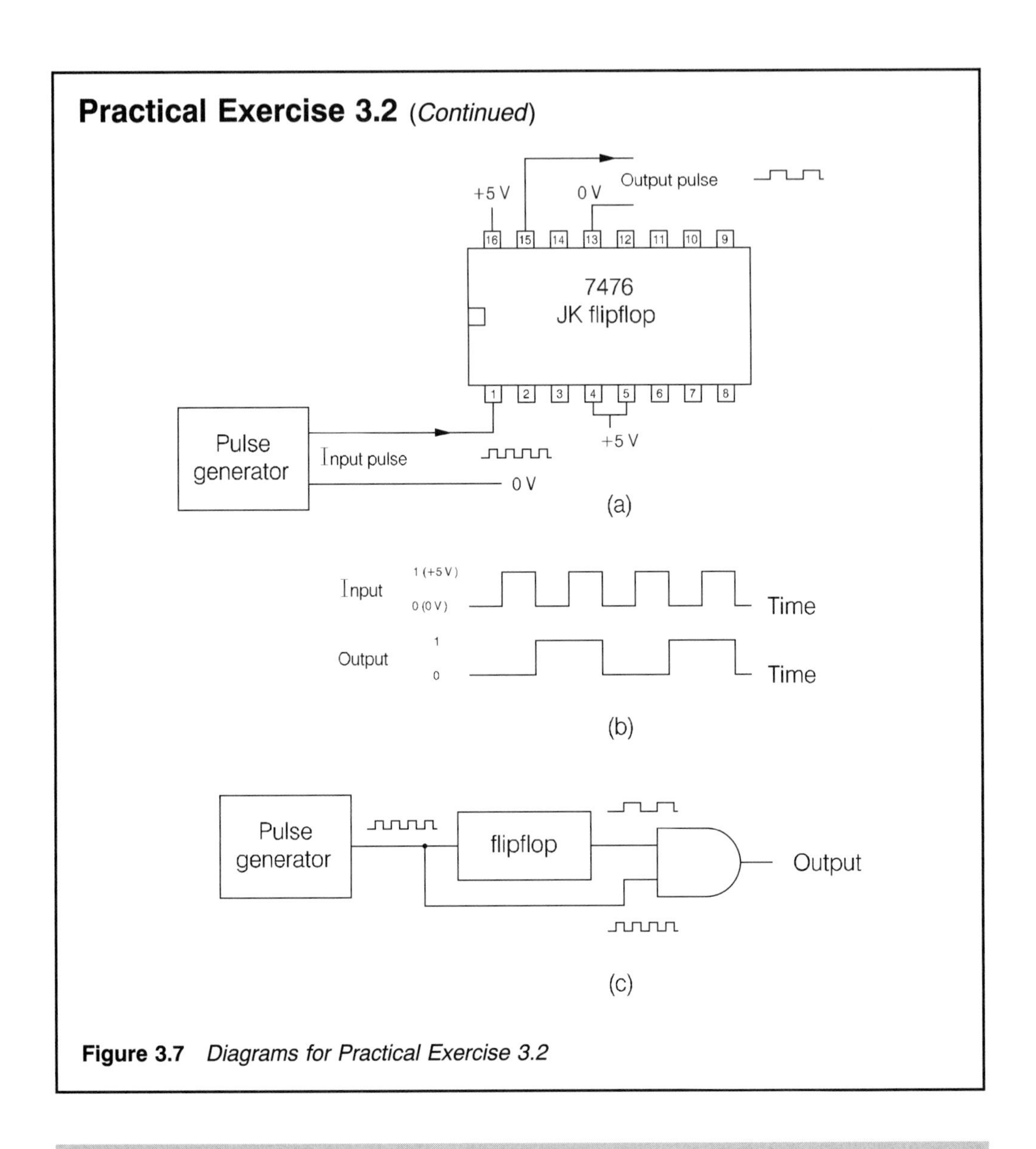

Figure 3.7 *Diagrams for Practical Exercise 3.2*

Questions

For each of the following questions it is suggested that you redraw the given waveforms on graph paper.

3.7 The waveforms shown in Figure 3.8 are applied to

(a) a 2-input AND gate
(b) a 2-input OR gate
(c) a 2-input NAND gate
(d) a 2-input NOR gate

Draw the output waveform of each gate.

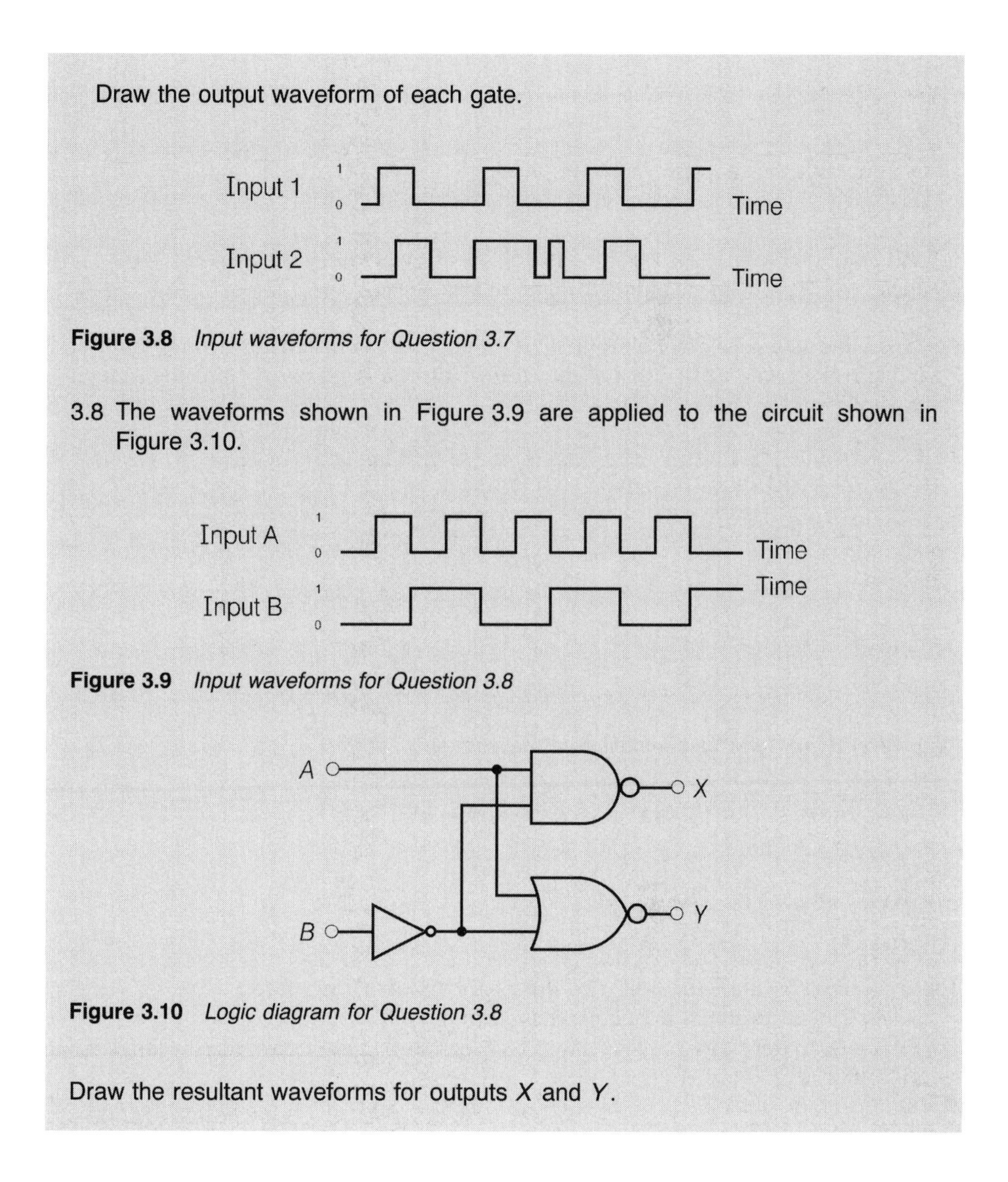

Figure 3.8 *Input waveforms for Question 3.7*

3.8 The waveforms shown in Figure 3.9 are applied to the circuit shown in Figure 3.10.

Figure 3.9 *Input waveforms for Question 3.8*

Figure 3.10 *Logic diagram for Question 3.8*

Draw the resultant waveforms for outputs X and Y.

The traffic light system

A good application for combinational logic problems is the well-known traffic light system. The arrangement is explained below, in preparation for the next practical exercise.

The traffic light sequence is

Red	(R)	followed by
Red and yellow	($R\&Y$)	followed by

Green (G) followed by
Yellow (Y) followed by
Red

and so on, with the sequence repeating itself.
At this stage it will be as well to make the assumptions that:

(a) the time interval for each of the four different lines above is the same;
(b) yellow LEDs are easier to obtain than amber!

This four-line sequence can be provided by a binary system having two inputs A and B, and three outputs R, Y and G. The truth table can be derived from the sequence given above, and is shown in Figure 3.11.

Inputs A B	Outputs R Y G	Logic expression
0 0	1 0 0	$R = \overline{A}.\overline{B}$
0 1	1 1 0	$R = \overline{A}.B$ $Y = \overline{A}.B$
1 0	0 0 1	$G = A.\overline{B}$
1 1	0 1 0	$Y = A.B$

Figure 3.11 *Truth table for the traffic light sequence*

The resulting logic expressions for R, Y and G are

$$R = \overline{A} \cdot \overline{B} + \overline{A} \cdot B$$

$$Y = \overline{A} \cdot B + A \cdot B$$

$$G = A \cdot \overline{B}$$

These can then be implemented with the use of suitable logic gates.

You may see that two of the above logic expressions can in fact be simplified. If it is not obvious, do not worry because later on you can use the established technique for simplification (see Chapter 4).

Practical Exercise 3.3

Traffic lights (1)
For this exercise you will need the following components and equipment:

1 – +5 V DC power supply
1 – set of logic gates as required
3 – LED (5 mm) (1 red, 1 yellow, 1 green and series resistors 270 Ω)

Procedure

1 Draw the logic diagrams using AND, OR and NOT gates as appropriate to show how the outputs R, Y and G can be implemented.
2 Connect up the system and test it.
3 Redraw the diagrams using NAND gates only.
4 Connect up and test your revised system.
5 Consult a catalogue to establish the minimum number of logic gates necessary to implement (1) and (3) above and the likely cost of each. Hence state, as a percentage, any financial benefit of (3).

4 Karnaugh mapping

We have seen in Chapter 3 that it is possible, by sometimes laborious methods, to simplify a logic output expression. The advantage in attempting to simplify will be apparent if we compare the logic diagrams of a system, before and after the simplification process.

Let us look again at the expression

$$Z = \overline{A} \cdot B + A \cdot \overline{B} + A \cdot B$$

The logic diagram for this is shown in Figure 4.1.

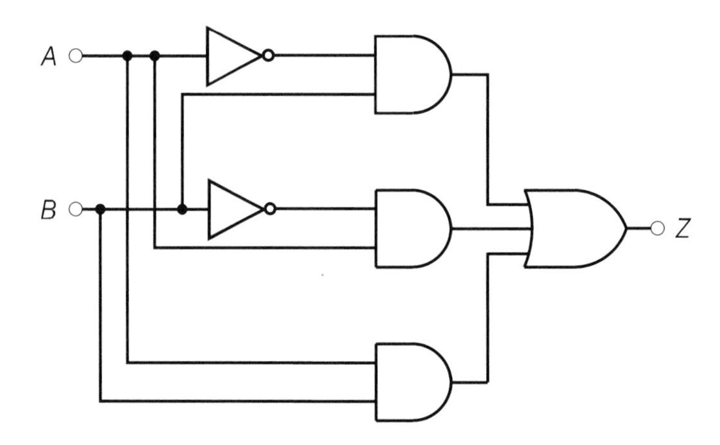

Figure 4.1 *Logic diagram to represent the expression* $Z + \overline{A} \cdot B + A \cdot \overline{B} + A \cdot B$

This function can be implemented, that is, made up, with two NOT gates, three (2-input) AND gates and one (3-input) OR gate. This gives a total of six gates. Of greater importance is the **number of chips**, which in this case amounts to three.

The simplified expression, $Z = A + B$ (which you will remember was the OR gate), requires just one ic chip, giving a saving of two ic's. For more involved expressions, the saving can be far greater, as we shall see as we continue working through this chapter.

Let us work one more example before we move into the Karnaugh mapping technique.

Example (a)

See Figure 4.2. Derive the logic statement for this system, draw up the truth table and, as a result, suggest an alternative logic circuit that could be used.

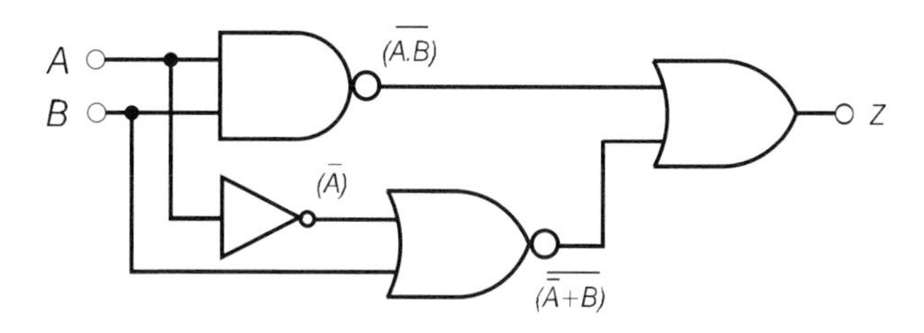

Figure 4.2 *Logic diagram for example (a)*

The logic statement is given by

$$Z = \overline{A \cdot B} + \overline{(\bar{A} + B)}$$

The truth table is shown in Figure 4.3.

A	B	$A.B$	$\overline{A.B}$	$\bar{A}$	$(\bar{A}+B)$	$\overline{(\bar{A}+B)}$	$\overline{A.B} + \overline{(\bar{A}+B)} = Z$
0	0	0	1	1	1	0	1
0	1	0	1	1	1	0	1
1	0	0	1	0	0	1	1
1	1	1	0	0	1	0	0

this is the output of a NAND gate

Figure 4.3 *Truth table for the system of Figure 4.2*

The column for Z will be recognized as the output of a NAND gate. Thus the simplification requires one NAND gate (= one ic), compared with four different gates (= four ic's) in the original expression.

Write down the output of each gate (as shown on the right-hand side of the gate).

Build up the truth table in stages as shown to the right-hand side of inputs A and B.

The Karnaugh Map (pronounced 'car-no')

The technique of Karnaugh mapping was developed in the early 1950s, as a means of attempting to simplify logic expressions.

The Karnaugh map is a graphical method of representing the truth table of a given logic function. The map is a rectangular diagram, the area of which is divided into squares, where the number of squares is the same as the number of lines in the truth table. (Remember that the number of lines in any truth table equals 2^n where n is the number of inputs. Thus for two inputs or variables there will four lines in the table, and hence four squares on the Karnaugh map.)

Each square on the map represents one particular line of the truth table.

The two-variable Karnaugh map

For an understanding of the principle of the Karnaugh map, look at Figure 4.4.

The procedure (referring to Figure 4.4) is:

(a) Draw a square and subdivide such that the number of small squares is either the same as the number of squares in the truth table or equal to 2^n as explained above.

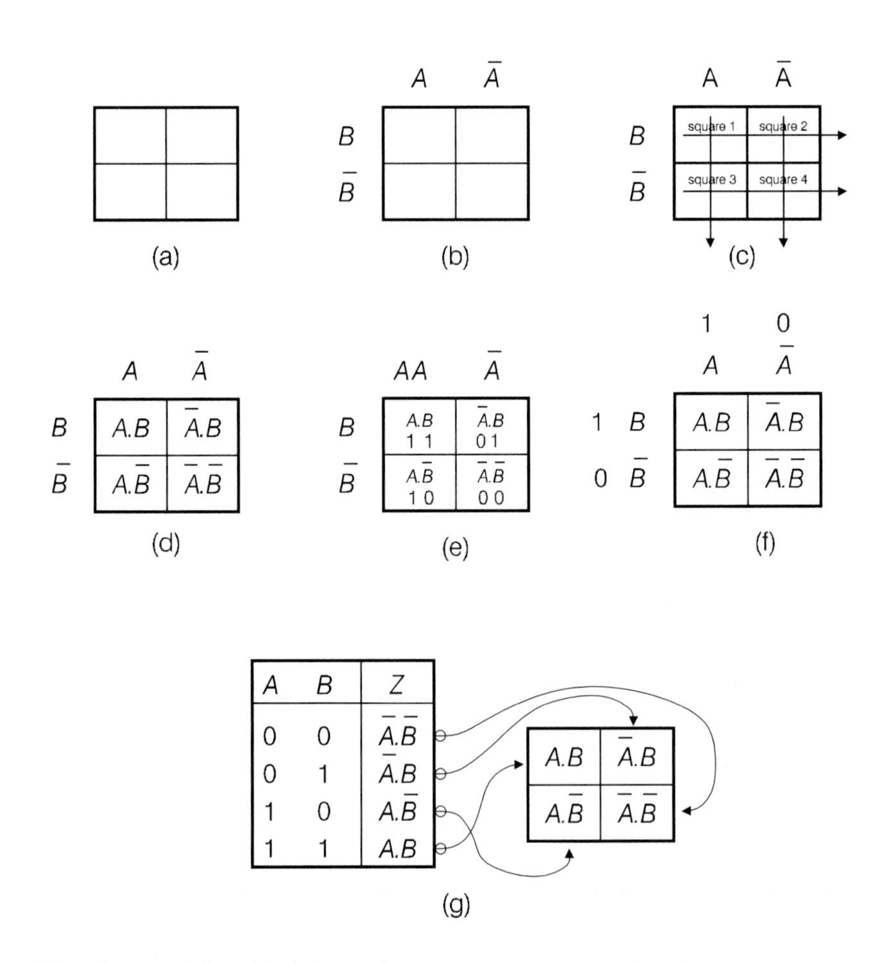

Figure 4.4 *The principle of the Karnaugh map*

(b) Letter the edges of the large square with variable A and its complement (opposite) $\overline{A}$ along the top, and variable B and its complement along the left-hand side.
(c) The identification of each square is by the intersection of the vertical and horizontal variable.
(d) Square 1 is thus given by $A \cdot B$; square 2 by $\overline{A} \cdot B$ and so on.

It is quite helpful:

(e) To write the binary grouping in each square; thus $A \cdot B = 11$, $\overline{A} \cdot B = 01$ and so on.
(f) To write the binary numbers for A, $\overline{A}$ etc. outside the large square.

Finally:

(g) Remember that each square on the Karnaugh map represents one line in the truth table.

An alternative way of arranging the map layout can be seen in Figure 4.5.

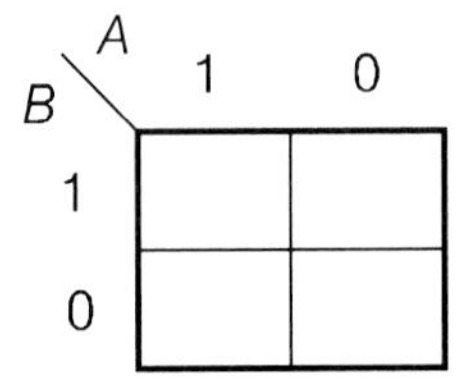

Figure 4.5 *Alternative layout for the Karnaugh map*

Important points

- The difference in the binary representation between squares 1 and 2 is by one digit – from 11 to 01.
  $A \cdot B$ goes to $\overline{A} \cdot B$, i.e. variable A changes but variable B does not. In other words, in working along the sides of the square, **one variable at a time** is changed to its complement.
- There is the same one-digit difference between squares 2 and 4, squares 4 and 3 and squares 3 and 1.
- Squares in which the binary representation differs by one digit are known as **adjacent squares**. The following squares are therefore *not* adjacent:

  1 and 4, 2 and 3.

To see how Karnaugh mapping is used, we will work some examples:

Example (b)

Map the function given by $Z = A \cdot B$.

The statement reads:

'Z will be 1 when A AND B are both 1'

The truth table for the function is shown in Figure 4.6 and the associated Karnaugh map in Figure 4.7. Note the connection between the 1 on the map and the bottom line for Z in the truth table. Do not expect any simplification here!

A	B	Z
0	0	0
0	1	0
1	0	0
1	1	1

Figure 4.6 *Truth table for* $Z = A \cdot B$

	A	$\overline{A}$
B	1	0
$\overline{B}$	0	0

No simplification possible here

Figure 4.7 *Karnaugh map for* $Z = A \cdot B$

Example (c)

Map the function given by $Z = A \cdot B + A \cdot \overline{B}$

Note that in this statement there are two (different) terms, namely $A \cdot B$ and $A \cdot \overline{B}$, and the statement for Z reads:

'Z will be 1 when A AND B are (both) 1, OR when A is 1 and B is NOT 1'

This will give two 1's in the truth table and two 1's on the K-map.

The map for the given function is shown in Figure 4.8.

	A	$\overline{A}$
B	1	0
$\overline{B}$	1	0

Figure 4.8 *Karnaugh map for* $Z = A \cdot B + A \cdot \overline{B}$

> **Important points**
>
> - The number of 1's in the truth table is the same as the number of (different) terms in the logic statement. It will be seen later on that there may be duplication of terms.
> - Again, there will be the same number of 1's on the map as there are 1's in the truth table.

A simplification

The technique in Karnaugh mapping is to draw a loop around the 1's in adjacent squares, as shown in Figure 4.9.

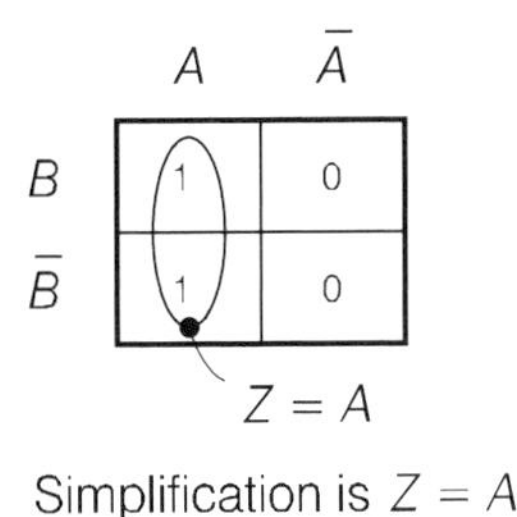

Figure 4.9 *Karnaugh map showing the looping for* $Z = A \cdot B + A \cdot \overline{B}$

The variable A does not change within the loop and remains. The variable B occurs as both $B = 1$ and $B = \overline{1}$ and can be eliminated. The reason is that since the variable B occurs as a 1 in both states (B and $\overline{B}$) the expression is therefore *independent* of the state of variable B. The simplification can thus be written as $Z = A$.

Drawing up the truth table (Figure 4.10) for the original function

$$Z = A \cdot B + A \cdot \overline{B}$$

will now show that this simplification is correct. The column for Z is identical to that for A.

> **Important points**
>
> - This (simple) example illustrates the use of the Karnaugh mapping technique, which is normally reserved for more involved expressions.
> - The function can also be simplified using the rules, or theorems:
>
> $$Z = A \cdot B + A \cdot \overline{B}$$

Continued on p. 46

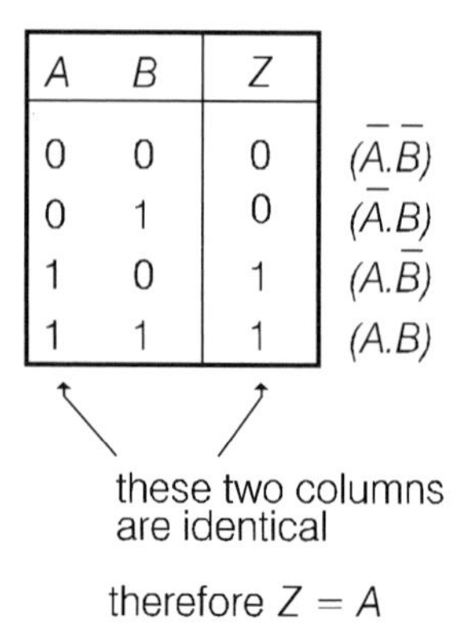

Figure 4.10 *Truth table for* $Z = A \cdot B + A \cdot \overline{B}$

Important points (*Continued*)

Factorizing gives

$$Z = A(B + \overline{B})$$

From theorem 10 (Figure 2.22),

$$(B + \overline{B}) = 1$$

Thus

$$Z = A \cdot 1$$

From theorem 12 (Figure 2.22),

$$A \cdot 1 = A$$

Thus, finally,

$$Z = A$$

Example (d)

Draw the Karnaugh map and use it to simplify the logic function

$$Z = \overline{A} \cdot B + \overline{A} \cdot \overline{B} + A \cdot B$$

There are three different components in this expression and there will be three 1's on the map (Figure 4.11).

Loops can be drawn around 1's in adjacent squares but must enclose 1's grouped in binary combinations, that is 2, 4, 8 and so on. In other words, a loop must contain 2, 4, 8 etc. 1's and *no other combination*.

A particular 1 may be used more than once in the looping process in order to produce the most simplified expression (known as the *minimal solution*).

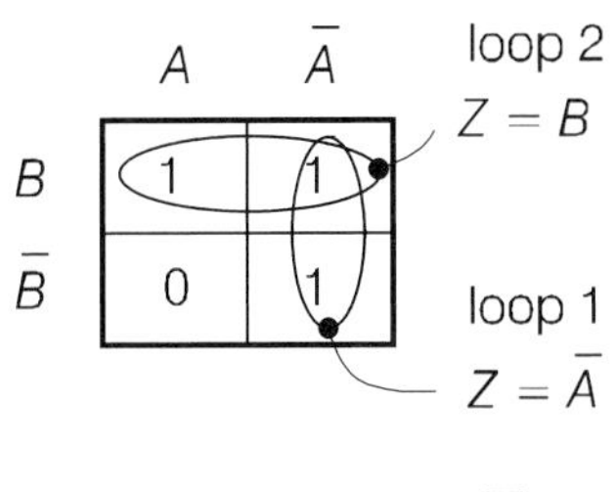

Simplification is $Z = \overline{A} + B$

Figure 4.11 *Karnaugh map for* $Z = \overline{A} \cdot B + \overline{A} \cdot \overline{B} + A \cdot B$

In this example, two loops can be drawn, that for $\overline{A} \cdot B$ being used twice. As a result, there are two parts of the simplification process:

(i) loop 1, which gives $Z = \overline{A}$, and
(ii) loop 2, which gives $Z = B$.

The simplified expression is given by

$$Z = \overline{A} + B$$

Important points

- The larger the loop, the greater the number of 1's inside this loop and the simpler the final expression will be.
- The Karnaugh map is a way of attempting to find a simplification, *but*
- Do not assume that a simplification is always possible. The next example should help to make this clear.

Example (e)

Attempt to simplify the expression $Z = A \cdot \overline{B} + \overline{A} \cdot B$

The Karnaugh map for this function is shown in Figure 4.12. The map shows that the two 1's are not in adjacent squares. Thus they cannot be looped and the expression cannot be simplified.

	A	$\overline{A}$
B	0	1
$\overline{B}$	1	0

The 1's are NOT in 'adjacent squares' and the expression cannot be simplified

Figure 4.12 *Karnaugh map for* $Z = A \cdot \overline{B} + \overline{A} \cdot B$

Question

4.1 Attempt a simplification of the following, using the Karnaugh map:

(a) $Z = \overline{A} \cdot B + A \cdot \overline{B} + A \cdot B$

(b) $Z = \overline{A} \cdot B + \overline{A} \cdot \overline{B} + A \cdot B$

(c) $Z = A \cdot B + \overline{A} \cdot \overline{B}$

The three-variable Karnaugh map

The three variables *A*, *B* and *C* will provide an eight-line truth table and hence an eight-square Karnaugh map. It is usual (but not obligatory) to keep the variables *A* and *B* together, which gives the map shown in Figure 4.13(a). Notice again how the squares in the map relate to the lines in the truth table, Figure 4.13(b).

		1 1	0 1	0 0	1 0
		$A.B$	$\overline{A}.B$	$\overline{A}.\overline{B}$	$A.\overline{B}$
1	C	$A.B.C$	$\overline{A}.B.C$	$\overline{A}.\overline{B}.C$	$A.\overline{B}.C$
0	$\overline{C}$	$A.B.\overline{C}$	$\overline{A}.B.\overline{C}$	$\overline{A}.\overline{B}.\overline{C}$	$A.\overline{B}.\overline{C}$

(a)

A	B	C	Z
0	0	0	$\overline{A}.\overline{B}.\overline{C}$
0	0	1	$\overline{A}.\overline{B}.C$
0	1	0	$\overline{A}.B.\overline{C}$
0	1	1	$\overline{A}.B.C$
1	0	0	$A.\overline{B}.\overline{C}$
1	0	1	$A.\overline{B}.C$
1	1	0	$A.B.\overline{C}$
1	1	1	$A.B.C$

(b)

Figure 4.13 *The 3-variable Karnaugh map*

Important points

- The squares represented by $A \cdot B \cdot C$ and $A \cdot \overline{B} \cdot C$ are adjacent, by definition, as are those represented by $A \cdot B \cdot \overline{C}$ and $A \cdot \overline{B} \cdot \overline{C}$. This is called **side-to-side adjacency**.
- The map can be bent to form a cylinder, in this case from side to side, but in other arrangements it could be from top to bottom.

Example (f)

Simplify the expression

$$Z = \overline{A} \cdot \overline{B} \cdot \overline{C} + \overline{A} \cdot B \cdot C + A \cdot B \cdot C \cdot + A \cdot \overline{B} \cdot \overline{C}$$

Figure 4.14 shows the map, which provides two loops. Loop 1 gives $Z = B \cdot C$ and loop 2 gives $Z = \overline{B} \cdot \overline{C}$, which combine to give the simplification

$$Z = B \cdot C + \overline{B} \cdot \overline{C}$$

(The simplification has eliminated variable A.)

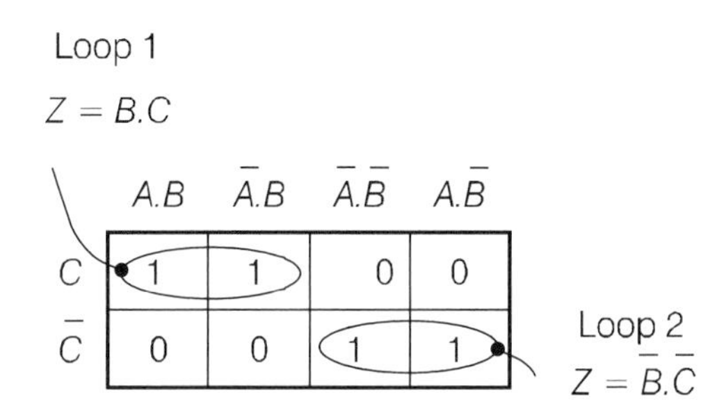

Figure 4.14 *The Karnaugh map for* $Z = \overline{A} \cdot \overline{B} \cdot \overline{C} + \overline{A} \cdot B \cdot C + A \cdot B \cdot C + A \cdot \overline{B} \cdot \overline{C}$

Important point

- The setting up of the map with the arrangement of the variables is a matter of personal choice, but having chosen, the rules regarding 'changing one variable at a time' must be obeyed.

  To help make this clear, example (f) is repeated, but with a slightly different mapping arrangement of variables, as shown in Figure 4.15.

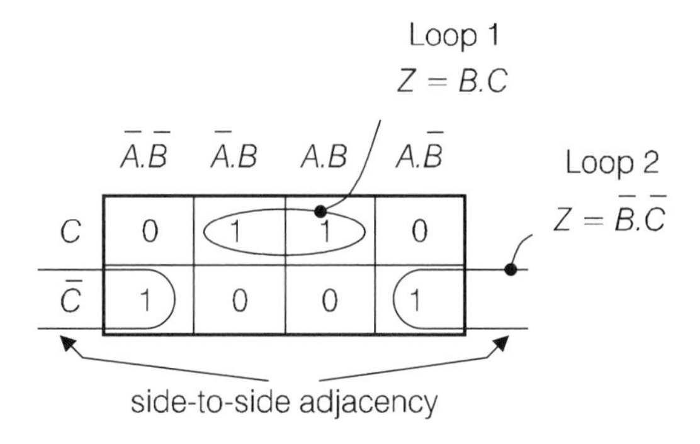

Figure 4.15 *An alternative Karnaugh map for* $Z = \overline{A} \cdot \overline{B} \cdot \overline{C} + \overline{A} \cdot B \cdot C + A \cdot B \cdot C + A \cdot \overline{B} \cdot \overline{C}$

Loop 1 gives

$$Z = B \cdot C$$

Continued on p. 50

Important point (*Continued*)

Loop 2, which is formed from side-to-side adjacency, gives

$$Z = \overline{B} \cdot \overline{C}$$

As a result (no different from before!),

$$Z = B \cdot C + \overline{B} \cdot \overline{C}$$

Example (g)

Simplify $Z = A \cdot B + A \cdot \overline{B} \cdot C + A \cdot B \cdot \overline{C}$

This is a three-variable expression, but the variables A, B and C are not all present in every part of the expression.

It is necessary to remember here that for the term $A \cdot B$ it is unimportant whether C is present or not. In other words, $A \cdot B$ started life as

$$A \cdot B \cdot C + A \cdot B \cdot \overline{C}$$

which can be written as

$$A \cdot B \cdot (C + \overline{C})$$

or, from theorem 10 (Figure 2.22), $(C + \overline{C} = 1)$ as

$$A \cdot B$$

It is now possible to rewrite the expression for Z as

$$Z = A \cdot B \cdot C + A \cdot B \cdot \overline{C} + A \cdot \overline{B} \cdot C + A \cdot B \cdot \overline{C}$$

Notice that there is a repeated term in this full expression for Z. As a result, there will be three 1's rather than four on the Karnaugh map.

The reader is invited to draw the map and, as a result, show that the simplification is given by

$$Z = A \cdot B + A \cdot C$$

Questions

4.2 Use the Karnaugh mapping technique to find the minimal (simplest) expression for each of the following examples.

Hint. One of the examples *will not* simplify.

(a) $Z = A \cdot \overline{B} \cdot C + \overline{A} \cdot B \cdot \overline{C}$
(b) $Z = A \cdot B \cdot C + A \cdot B \cdot \overline{C}$
(c) $Z = \overline{A} \cdot B \cdot C + A \cdot B \cdot \overline{C} + A \cdot B \cdot C$
(d) $Z = \overline{A} \cdot B \cdot C + A \cdot \overline{C} + A \cdot B \cdot C$
(e) $Z = A \cdot B \cdot \overline{C} + \overline{A} \cdot \overline{B} \cdot C + A \cdot B \cdot C + A \cdot \overline{B} \cdot C$

(f) $Z = A \cdot B \cdot C + A \cdot \overline{B} \cdot \overline{C} + \overline{A} \cdot \overline{B} \cdot \overline{C}$
(g) $Z = A \cdot B \cdot C + \overline{A} \cdot B \cdot \overline{C} + A \cdot B \cdot \overline{C} + \overline{A} \cdot B \cdot C$

4.3 From the truth tables in Figure 4.16 write down the logic expression for each. Use the Karnaugh mapping technique to find the minimal solution for each expression.

(a)

A	B	C	Z
0	0	0	0
0	0	1	0
0	1	0	0
0	1	1	0
1	0	0	1
1	0	1	0
1	1	0	1
1	1	1	0

(b)

A	B	C	Z
0	0	0	0
0	0	1	1
0	1	0	1
0	1	1	0
1	0	0	1
1	0	1	0
1	1	0	1
1	1	1	0

Figure 4.16 *Truth tables for Question 4.3*

The four-variable Karnaugh map

Karnaugh mapping comes into its own at this stage, the map consisting of $2^4 (= 16)$ squares. As with the earlier examples, the arrangement of the variables around the perimeter of the map is a matter of choice, but, being conventional, we shall keep together variables A and B and variables C and D.

Example (h)

Simplify

$$Z = \overline{A} \cdot \overline{B} \cdot \overline{C} \cdot D + A \cdot B \cdot \overline{C} \cdot D + A \cdot \overline{B} \cdot \overline{C} \cdot D + \overline{A} \cdot \overline{B} \cdot C \cdot D + A \cdot \overline{B} \cdot C \cdot D.$$

The map is shown in Figure 4.17.

Loop 1, enclosing the group of four 1's, gives

$$Z = \overline{B} \cdot D$$

Loop 2, formed by side-to-side adjacency, gives

$$Z = A \cdot \overline{C} \cdot D$$

The minimal (simplest) expression is given by

$$Z = \overline{B} \cdot D + A \cdot \overline{C} \cdot D$$

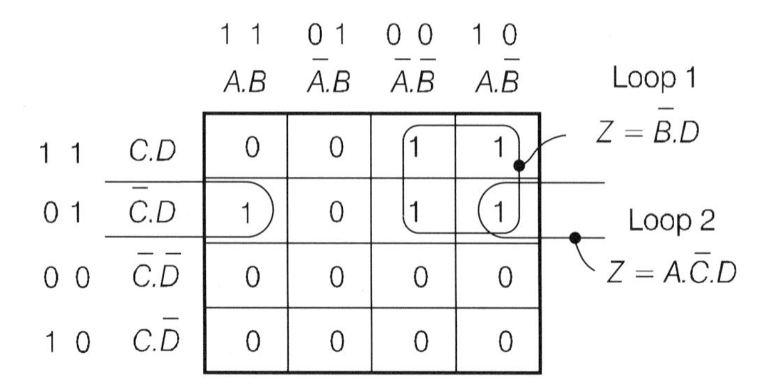

Figure 4.17 *The Karnaugh map for* $Z = \overline{A}\cdot\overline{B}\cdot\overline{C}\cdot D + A\cdot B\cdot\overline{C}\cdot D + A\cdot\overline{B}\cdot\overline{C}\cdot D + \overline{A}\cdot\overline{B}\cdot C\cdot D + A\cdot\overline{B}\cdot C\cdot D$

You should now draw the Karnaugh map using a different arrangement of the variables and arrive at the same solution.

Questions

Use a Karnaugh map in each of the following:

4.4 $Z = \overline{A}\cdot B\cdot\overline{C}\cdot\overline{D} + \overline{A}\cdot\overline{B}\cdot\overline{C}\cdot D + \overline{A}\cdot\overline{B}\cdot C\cdot D + \overline{A}\cdot B\cdot C\cdot\overline{D} + A\cdot\overline{B}\cdot\overline{C}\cdot D + A\cdot\overline{B}\cdot C\cdot D$

4.5 $Z = \overline{A}\cdot B\cdot\overline{C}\cdot\overline{D} + A\cdot B\cdot\overline{C}\cdot\overline{D} + \overline{A}\cdot B\cdot\overline{C}\cdot D + A\cdot B\cdot\overline{C}\cdot D + A\cdot\overline{B}\cdot C\cdot D + A\cdot\overline{B}\cdot C\cdot\overline{D}$

4.6 $Z = A\cdot\overline{B}\cdot\overline{C}\cdot\overline{D} + \overline{A}\cdot B\cdot\overline{C}\cdot D + \overline{A}\cdot\overline{B}\cdot\overline{C}\cdot D + \overline{A}\cdot\overline{B}\cdot C\cdot D + \overline{A}\cdot B\cdot C\cdot D + A\cdot\overline{B}\cdot\overline{C}\cdot D$

4.7 From the truth table in Figure 4.18, and the use of a Karnaugh map, find the simplest form of the expression for Z.

A	B	C	D	Z
0	0	0	0	1
0	0	0	1	1
0	0	1	0	1
0	0	1	1	1
1	1	1	0	1
1	1	1	1	1

All other combinations of the input variables give $Z = 0$

Figure 4.18 *Truth table for Question 4.7*

Don't care/can't happen states

For some practical logical circuits there are particular combinations of input variables which cannot occur (the **can't happen** state). The output for this combination cannot be given and indeed, is unimportant. For other circuits there may be combinations which may or may not occur (the **don't care** state). Again, the outcome (whether a 1 or a 0) is unimportant.

As an example, consider the situation where there are three switches A, B and C, and only when two of them, say A and B, are on together will there be a recognized output. In other words, the logical situation is given by

$$Z = A \cdot B \cdot \overline{C}$$

or

$$A \cdot B \cdot C = 110.$$

The truth table is shown in Figure 4.19. The last line of the truth table, $ABC = 111$, is a 'can't happen' state, since it was stated at the beginning that only two switches could be on at any one time. The letter X is used to denote a 'can't happen' state and the Karnaugh map is shown in Figure 4.20.

A	B	C	Z
0	0	0	0
0	0	1	0
0	1	0	0
0	1	1	0
1	0	0	0
1	0	1	0
1	1	0	1
1	1	1	X

Figure 4.19 *Truth table for 'can't happen' state* $Z = A \cdot B \cdot \overline{C}$

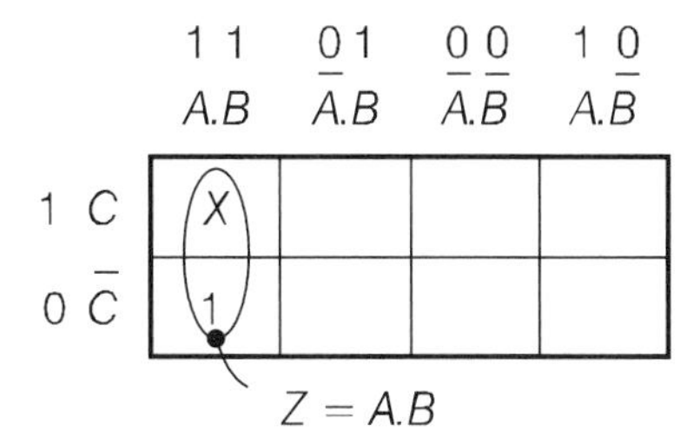

Figure 4.20 *The Karnaugh map for* $Z = A \cdot B \cdot \overline{C}$

Notice that the 1 and X are in adjacent squares (110 and 111 are adjacent by definition). When an X is shown on the map it can be taken to be either a 1 or a 0, whichever is more appropriate for the simplification process. Thus the X and 1 can be looped, as shown, to provide the minimal expression. The loop gives the result $Z = A \cdot B$.

Important point

- There may be several X's on the map, but it is not necessary to include them all in the looping process. In any case, it may not be possible to do so. Only the X's that give minimization need be used.

Example (i)

A 'can't happen' state occurs, for example, in a binary coded decimal (BCD) system, where the four bits A, B, C and D represent the decimal numbers 0 to 9. The binary equivalents for the decimal numbers 10 to 15 cannot occur.

Suppose that the design requirement is for the output (Z) of a system to go high when the decimal input is 2, 4, 6 or 8. The (partial) truth table is shown in Figure 4.21. The four output 1's, corresponding to 0010, 0100, 0110 and 1000, are marked on the Karnaugh map, together with the six X's representing the 'can't happen' states. It is also useful to mark on the six 0's. See Figure 4.22.

decimal	A B C D	Z	
2	0 0 1 0	1	$(\bar{A}.\bar{B}.C.\bar{D})$
4	0 1 0 0	1	$(\bar{A}.B.\bar{C}.\bar{D})$
6	0 1 1 0	1	$(\bar{A}.B.C.\bar{D})$
8	1 0 0 0	1	$(A.\bar{B}.\bar{C}.\bar{D})$
10	1 0 1 0	X	$(A.\bar{B}.C.\bar{D})$
11	1 0 1 1	X	$(A.\bar{B}.C.D)$
12	1 1 0 0	X	$(A.B.\bar{C}.\bar{D})$
13	1 1 0 1	X	$(A.B.\bar{C}.D)$
14	1 1 1 0	X	$(A.B.C.\bar{D})$
15	1 1 1 1	X	$(A.B.C.D)$

All other combinations of the input variables give $Z = 0$

Figure 4.21 *Truth table for example (i)*

Figure 4.23 (a) to (d) shows how different looping arrangements can be used in the simplification process. The expression for Z found by looping the 1's only is given by

$$Z = A \cdot \bar{B} \cdot \bar{C} \cdot \bar{D} + \bar{A} \cdot C \cdot \bar{D} + \bar{A} \cdot B \cdot \bar{D}$$

	1 1 $A.B$	0 1 $\overline{A}.B$	0 0 $\overline{A}.\overline{B}$	1 0 $A.\overline{B}$
1 1 $C.D$	X	0	0	X
0 1 $\overline{C}.D$	X	0	0	0
0 0 $\overline{C}.\overline{D}$	X	1	0	1
1 0 $C.\overline{D}$	X	1	1	X

Figure 4.22 *The Karnaugh map for example (i)*

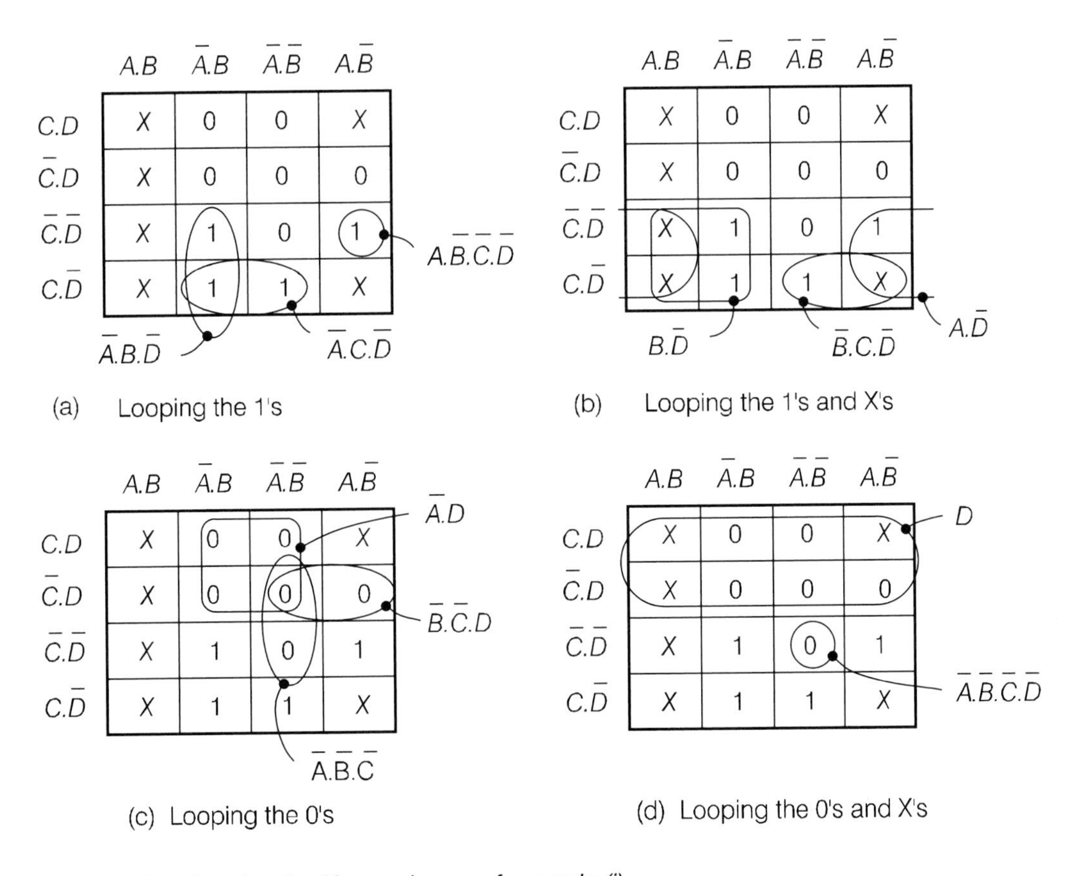

Figure 4.23 *Looping the Karnaugh map of example (i)*

To help the simplification we can loop the 1's and X's, and get

$$Z = A \cdot \overline{D} + B \cdot \overline{D} + \overline{B} \cdot C \cdot \overline{D}$$

Looping the 0's only gives the inverse expression, that is $\overline{Z}$, given by

$$\overline{Z} = \overline{A} \cdot D + \overline{B} \cdot \overline{C} \cdot D + \overline{A} \cdot \overline{B} \cdot \overline{C}$$

This is the expression for $Z = 0$, and simplification may be possible by looping the 0's and X's, following which we get

$$\overline{Z} = \overline{A} \cdot \overline{B} \cdot \overline{C} \cdot \overline{D} + D$$

Practical Exercise 4.1

Traffic lights (2)
For this exercise you will need the following components and equipment:

1 – +5 V DC power supply
1 – set of logic gates as required
3 – LED (5 mm) (1 red, 1 yellow, 1 green and series resistors 270 Ω)

Since this exercise carries on where Practical Exercise 3.3 left off, you will need to refer back and write down here the logic expressions for R, Y and G.

Procedure

1. Use the Karnaugh mapping technique (or the rules of Boolean algebra!) to attempt a simplification of the logic expressions. Draw the revised logic diagram.
2. Connect up and test your revised system.
3. Convert the revised diagram into NAND gates and repeat the costing exercise performed earlier.
   In the above arrangement, the lights spent the same amount of time on each colour or combination of colours.
   The system is now to be altered such that the time for either red or green separately is three times longer than for yellow or red and yellow. In other words, the sequence becomes:

   red, red, red, red and yellow, green, green, green, yellow

   The total time interval adds up to eight units-worth. This sequence can therefore be achieved with an eight-line truth table with, of course, three inputs.
4. Draw up this truth table and as a result obtain the new logic expressions for R, Y and G. (The expression for R will contain four terms, that for Y will have two terms and that for G will have three terms. They will all benefit from an attempted simplification.)
5. Use Karnaugh mapping to provide a simplification for all three outputs and draw the logic diagram (with AND, OR and NOT gates as before).
6. Connect up and test this system.
   Taking the arrangement a stage further, the time for red and green separately is now to be five times that for yellow or red and yellow.
7. Decide on the total time interval and hence the number of lines in the truth table, together with the number of inputs necessary to achieve the sequence.

8 Draw up the truth table and obtain the expressions for *R*, *Y* and *G* (AND, OR and NOT gates).

9 Use Karnaugh mapping to attempt a simplification for each output. (All outputs will simplify. Remember: if you have more lines in the truth table than are necessary, the remainder should be included in the Karnaugh map as 'don't cares'.)

10 Connect up and test this system. *Hint.* If you need three-input AND and OR gates, remember that these can be made up from two-input gates.

11 Perform one more costing exercise, by comparing the implementation of (10) above with the equivalent in NAND gates.

Static hazards

A logic gate takes a finite time to operate, of the order of milliseconds, with not all the gates necessarily having this same time. As a result, errors may arise in the output of the circuit. For example, look at the Karnaugh map of Figure 4.24 which has produced the simplified expression

$$Z = A \cdot B + \overline{A} \cdot C$$

The logic diagram is shown in Figure 4.25, together with the logic state of each gate with inputs $ABC = 111$, and the truth table is given in Figure 4.26.

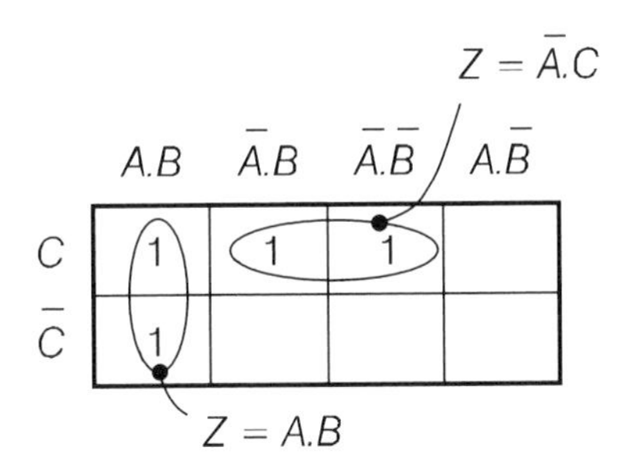

Figure 4.24 *The Karnaugh map giving the simplified expression* $Z = A \cdot B + \overline{A} \cdot C$

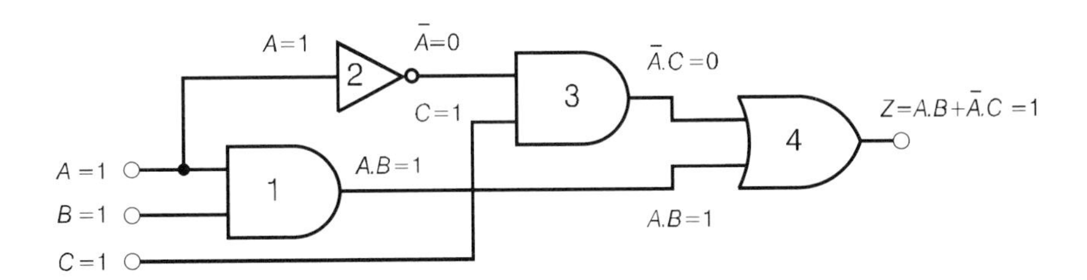

Figure 4.25 *Logic diagram for* $Z = A \cdot B + \overline{A} \cdot C$

A	B	C	$\bar{A}$	$\bar{A}.C$	$A.B$	Z
0	0	0	1	0	0	0
0	0	1	1	1	0	1
0	1	0	1	0	0	0
0	1	1	1	1	0	1
1	0	0	0	0	0	0
1	0	1	0	0	0	0
1	1	0	0	0	1	1
1	1	1	0	0	1	1

Figure 4.26 *Truth table for* $Z = A \cdot B + \bar{A} \cdot C$

The initial conditions are $ABC = 111$, giving (from the truth table) $Z = 1$.

Suppose now that it is necessary for the conditions to change to $ABC = 011$, which should provide no change in the output Z (see truth table to confirm this). If, however, there is a longer delay in the NOT gate, then the sequence is as follows:

- Input A changes from 1 to 0
- Output of gate 1 (AND) changes from 1 to 0
- Output of gate 2 (NOT) is delayed, therefore remains at 0 (should have changed to 1)
- Output of gate 3 (AND) remains at 0 (should have changed to 1)
- Output of gate 4 (OR), the output of the system, changes from 1 to 0 (should have remained unchanged at 1)

This change in the output state should not have occurred and is known as a **transient change** or **glitch**. The condition is regarded as a **static hazard**. The practical application of this means that if a change of output from 1 to 0 is normally intended to trigger an alarm, then the presence of the hazard will cause an unintentional alarm signal.

Figure 4.27 shows the logic levels following the change of input, while Figure 4.28 illustrates the action by means of waveform diagrams.

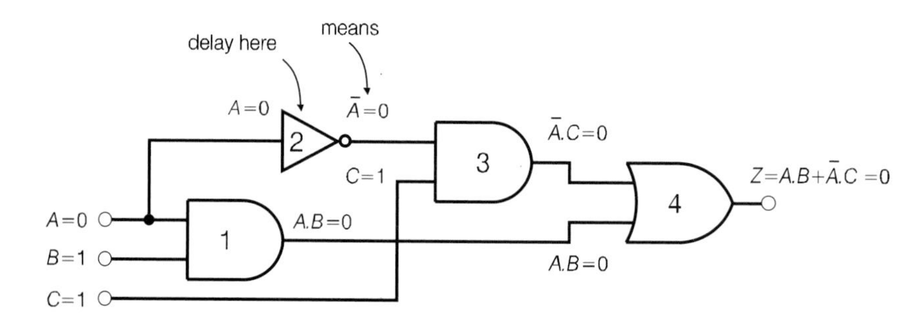

Figure 4.27 *Implementation of* $Z = A \cdot B + \bar{A} \cdot C$ *following a change in ABC, and a delay in gate 2*

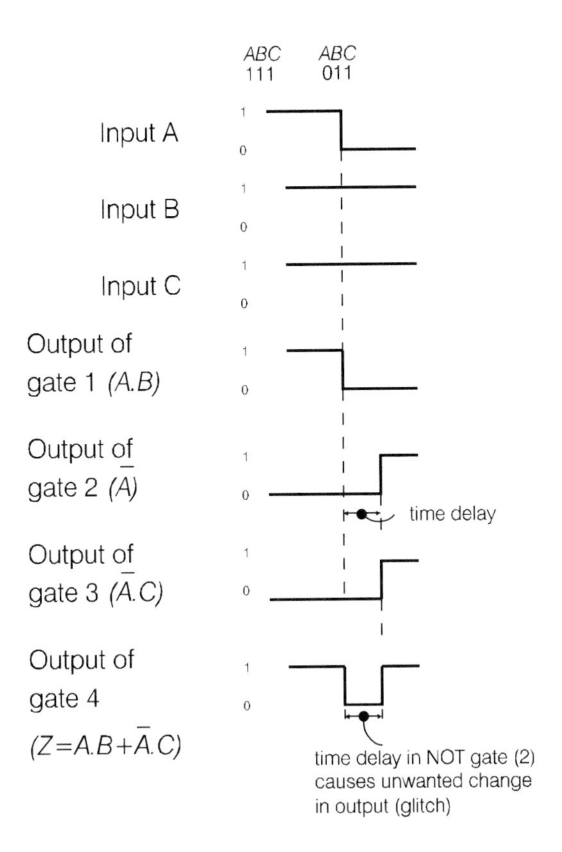

Figure 4.28 *Waveforms showing the effect of a static hazard*

The hazard is likely to arise if there are unlooped 1's in adjacent squares on the Karnaugh map. Hence, to remove this likelihood, the technique is to ensure that all adjacent 1's are looped together. The Karnaugh map of Figure 4.24 is repeated in Figure 4.29 with the appropriate modifications.

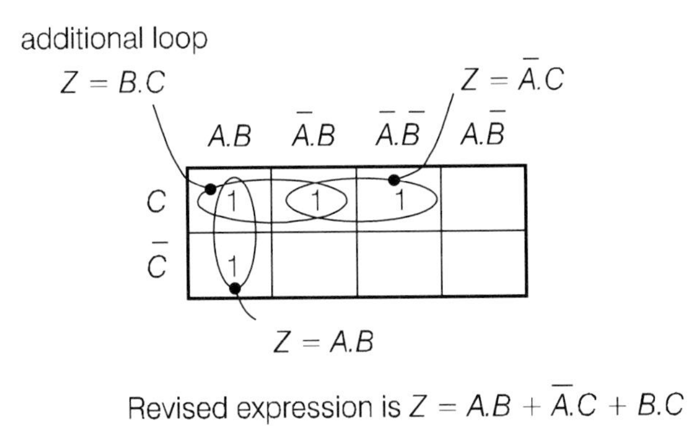

Figure 4.29 *The Karnaugh map showing additional term necessary to remove the static hazard*

The additional loop will provide an additional term in the expression for Z and hence an additional gate in the implementation. This gate, in pure logic terms, is redundant, but its presence ensures that the change of input will not cause the momentary glitch.

The revised logic circuit is shown in Figure 4.30 and its truth table in Figure 4.31.

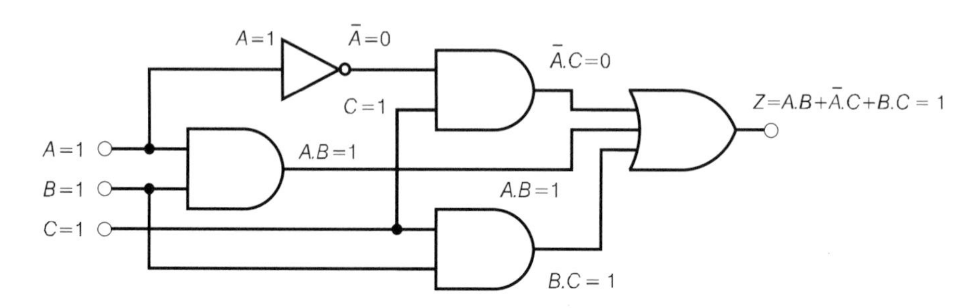

Figure 4.30 *Implementation of* $Z = A \cdot B + \overline{A} \cdot C + B \cdot C$ *with* $ABC = 111$

A	B	C	$\overline{A}$	$\overline{A}.C$	A.B	B.C	Z
0	0	0	1	0	0	0	0
0	0	1	1	1	0	0	1
0	1	0	1	0	0	0	0
0	1	1	1	1	0	1	1
1	0	0	0	0	0	0	0
1	0	1	0	0	0	0	0
1	1	0	0	0	1	0	1
1	1	1	0	0	1	1	1

Figure 4.31 *Truth table for* $Z = A \cdot B + \overline{A} \cdot C + B \cdot C$

Important point

- The line for $B \cdot C$ in the truth table (Figure 4.31) is a 1 for both input conditions $ABC = 111$ and $ABC = 011$, thus ensuring that one input to the OR (output) gate will always be a 1 while this particular input change is taking place.

Questions

4.8 A particular logic system has four binary inputs, *A*, *B*, *C* and *D*, which go through the decimal sequence 0 to 9. The output is required to go high for decimal inputs of 1, 2, 3, 5 and 7. Design a suitable logic circuit showing the additional term necessary for the system to be free of static hazards. *Remember*: decimal numbers 10 to 15 cannot occur.

4.9 A logic circuit is required which provides a logic 1 output when any two (but not three) of its three inputs is logic 1. Devise an appropriate circuit.
Make the necessary modifications to include the situation whereby the output is 1 when all three inputs are 1.
In both cases, attempt to find the minimal expression.

5 Logic families

Digital integrated circuits have been developed and manufactured as **families**. These families are groups of devices that are intended to be used together. The different types of ic within any one family are compatible and can be easily connected together.

It is possible for digital ic's from different families to be connected together, but special interfacing techniques must be used.

The three types of family

TTL – transistor transistor logic

This family uses bipolar technology, that is to say, the fabrication of the chip is in terms of resistors, capacitors, diodes and bipolar transistors. (The term **bipolar** means that the current within the device is due to the movement of both holes and electrons.)

For satisfactory switching, the transistors have to be fully saturated. This will put a limit on the speed of operation.

The TTL family is identified in its commercial form as the 74XX series and is available in several versions. Taking the quad two-input NAND gate (7400) as an example, the various types are listed in Figure 5.1. The military version in which more stringent conditions have to be met, is known as the 5400 series.

The TTL family is a very popular and widely used family in one or other of its forms.

Type	Designation	Advantage over 'Standard' type
Standard TTL	7400	
Schottky TTL (S)	74S00	Faster-acting. (Superseded)
Low Power Schottky TTL (LS)	74LS00	Lower power
Advanced Low Power Schottky TTL (ALS)	74ALS00	Faster and lower power
Fast (F)	74F00	Even faster

Figure 5.1 *Brief details of the TTL family*

CMOS – complementary metal oxide semiconductor

This family uses unipolar technology, as in field effect transistors, where current is by the movement of one type of charge carrier only (either holes or electrons). The CMOS type can be made using either type of transistor in its construction, hence the term 'complementary'.

The principal advantage of CMOS over TTL is its low power consumption which makes it ideal for portable battery-operated equipment.

CMOS is available in the 4XXX series for low power operation, at slower speed, or as a high-speed device in the 74HC or 74AC series.

ECL – emitter coupled logic

Bipolar technology is also used for this family of logic devices, but the transistors are operated in the unsaturated state. This makes ECL the fastest of all logic ic's, but at a higher power dissipation.

We have seen that the manufacture or fabrication of integrated circuits has involved both bipolar and metal oxide semiconductor technology. An important aspect of the technology is the number of gates and components that can be fabricated on a single chip. Integrated circuits can be classified on a scale of integration, in the following (generalized) manner:

Small-scale integration (SSI) – where the number of components per chip is less than 100.

Medium-scale integration (MSI) – number of devices is between 100 and 1000.

Large-scale integration (LSI) – number of devices is between 1000 and 10 000.

Very-large scale integration (VLSI) – number of devices is greater than 10 000.

VLSI requires the use of MOS technology, but the other three can use either bipolar or MOS.

Characteristics of logic families

What characteristics are used when choosing a logic gate?

You will see from what follows that it is a matter of compromise, usually between **speed of operation** and **power consumption** of a gate. The point to bear in mind with regard to power consumption is that although the power dissipated by a single gate may only be of the order of milliwatts, the need for perhaps several thousand such gates in a particular unit clearly raises the issue of whether or not that unit can be 'battery driven'.

Explanations are now given for the various characteristics, following which a table (Figure 5.13) provides a summary of typical values.

Supply voltage

Device	Supply voltage (V_{CC})	Comments
TTL	+5 V +/− 0.25 V	Stabilized PSU required
CMOS	+3 V to +15 V	
ECL	−5.2 V (the '10 000' series) −4.5 V (the '100 000' series)	

Logic levels

There are two types of logic – positive and negative.

Positive logic is where the more positive voltage gives logic 1 and the more negative voltage gives logic 0.

Negative logic, on the other hand, is where logic 1 is provided by the more negative voltage.

This is illustrated by the diagrams in Figure 5.2.

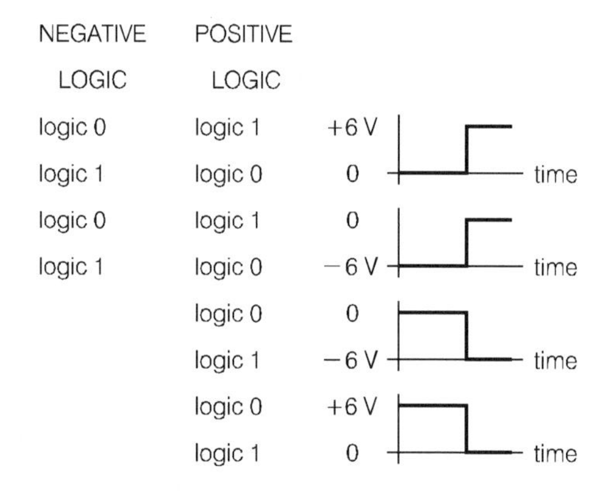

NEGATIVE LOGIC	POSITIVE LOGIC	
logic 0	logic 1	$+V_{CC}$
logic 1	logic 0	0 V
logic 0	logic 1	0 V
logic 1	logic 0	$-V_{CC}$

Figure 5.2 *Positive and negative logic*

You may think, taking TTL as the example, that in order to provide logic levels of 1 and 0 it is necessary to apply DC voltages of +5 V and 0 V respectively. In practice, however, there will be a range of voltages for which these logic levels will be achieved or, as we say, 'recognized'. The term 'logic level' refers to the input and output voltages that must be applied for the gate to operate correctly.

For TTL, which requires a power supply (V_{CC}) of +5 V and where, for positive logic, 0 V = logic 0 and +5 V = logic 1, the following situation applies:

At the input

(a) The minimum DC value for logic 1 = +2 V. In other words, a logic 1 will be registered at the gate input for voltages between +2 V and +5 V.
This is called the high level input voltage (V_{IH}). From the data in Figure 5.3, notice that V_{IHmin} = +2 V and V_{IHmax} = +5.5 V.

(b) The maximum DC value for logic 0 = +0.8 V. This means that logic 0 will be registered at the input for voltages between 0 V and +0.8 V.
This is the low level input voltage (V_{IL}). The data gives values of V_{ILmin} = 0 V and V_{ILmax} = +0.8 V. There is thus a voltage gap or 'band' from +0.8 V (V_{ILmax}) to +2 V (V_{IHmin}) a kind of 'no-man's land' where the logic level will be indeterminate. This situation *must* be avoided.

At the output

(a) The minimum DC value for logic 1 = +2.4 V. This means that logic 1 will be recognized at the output for voltages between +2.4 V and the permitted maximum.
This is the high level output voltage (V_{OH}). The data gives V_{OHmin} = +2.4 V and V_{OHmax} = +3.4 V.

(b) The maximum DC value for logic 0 = +0.4 V. This is the low level output voltage (V_{OL}). The data gives V_{OLmin} = +0.2 V and V_{OLmax} = +0.4 V.

Values for the TTL logic levels together with the appropriate abbreviations are shown in Figure 5.3. For CMOS logic, the level depend on the supply voltage (V_{DD}), which for successful operation can be from +3 V to +15 V. As far as the input is concerned, the maximum voltage for logic 0 is 0.3 V_{DD}, while the minimum voltage for logic 1 is 0.7 V_{DD}.

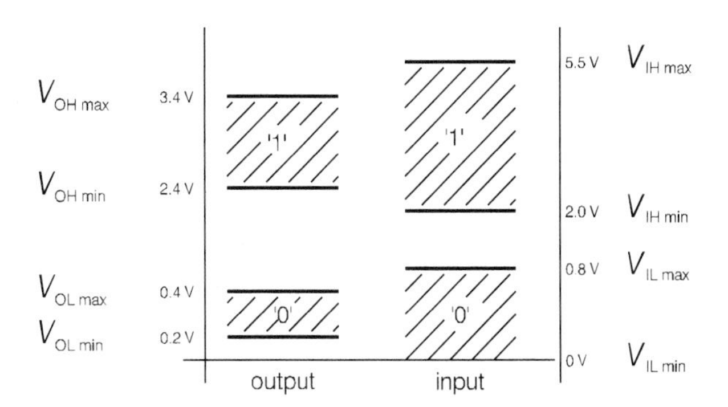

Figure 5.3 *Logic levels*

For ECL, logic 1 is represented by −0.8 V and logic 0 by −1.6 V.

Propagation delay

Reference has already been made to the speed of operation of digital circuits as a characteristic to be considered when choosing devices for particular purposes.

Let us take a brief look at the operation of the bipolar transistor as a switch, whose circuit is shown in Figure 5.4. With $V_{in} = 0$ V, there will be zero base current and the only current in the collector circuit will be that due to leakage. At this stage V_{CE} will be equal to V_{CC}.

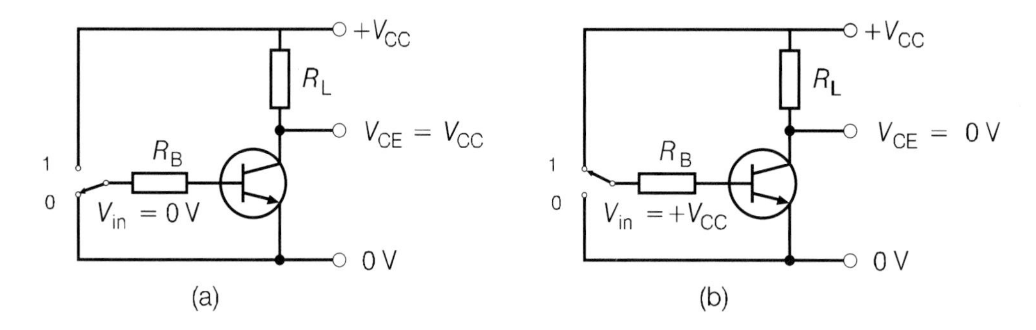

Figure 5.4 *The basic transistor switch*

For effective switching action, as V_{in} changes from 0 V to $+V_{CC}$, the transistor conducts (switches 'on') and V_{CE}, given by $V_{CC} - I_C R_L$ should become 0 V or, in practice, +0.2 V. This involves driving the transistor into its saturated state, with maximum collector current (I_{Csat}). There is a finite time delay while this happens, largely due to the charging time of the base-emitter capacitance. Any further increase in base current once the transistor has saturated must be stored in the base region.

This **charge storage** now causes a problem when, following a change in base voltage to zero, the transistor tries to switch off. The result is another delay while the excess charge is removed, and the collector voltage begins its return to $+V_{CC}$. The diagram in Figure 5.5 illustrates this effect.

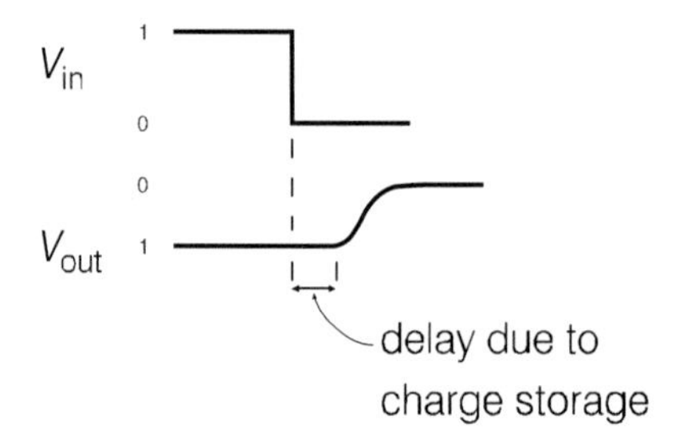

Figure 5.5 *The effect of charge storage*

Important points

- The practical square or rectangular wave is rarely 'ideal', that is, possessing vertical edges. There will always be a finite time for the pulse to go from zero to its maximum value (or from 10% to 90% of its final value). Consequently, the measurements in pulse work normally involve the 50% of the amplitude (or halfway) **line**.

  Propagation delay is the time taken between the application of the input signal and the resulting change in the output state.
- The time taken for the transistor to saturate, that is, for the output collector-emitter voltage V_{CE} to go from $+V_{CC}$ to effectively zero, is given as t_{pHL}.

  A typical value for t_{pHL} (for TTL 7400) is 7 ns.
- The time taken for the transistor to return to its unsaturated state is given as t_{pLH}. A typical value for t_{pLH} (TTL 7400) is 11 ns.
- These times, while close in numerical value, are not identical and manufacturers quote an average value t_{pD}, where

  $$t_{pD} = \frac{1}{2}(t_{pHL} + t_{pLH})$$
- Smaller propagation delays are available with unsaturated bipolar transistors, using Schottky diode techniques (3 ns), certain CMOS types (5 ns) and ECL logic (2 ns).

  Propagation delay is illustrated by the diagram in Figure 5.6.

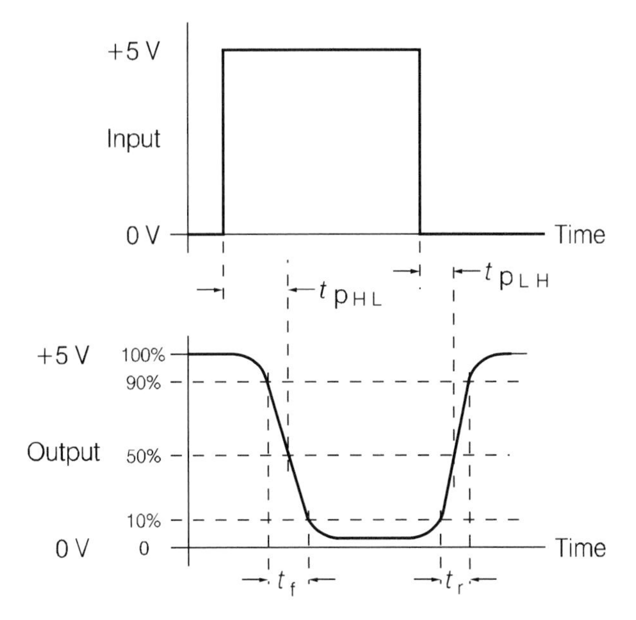

Figure 5.6 *Propagation delay of a logic gate*

Fan-in and fan-out

The **fan-in** (see Figure 5.7(a)) of a logic gate is the maximum number of inputs, from similar circuits, that can be connected to the gate without affecting its performance, that is, without causing the input logic level (voltage) to fall outside its specified range of values. One factor determining the fan-in is **switching speed**, since each input contributes a certain capacitance, which itself sets the limit on the propagation times.

Another, no less important but basic, factor is the number of pins available on the integrated package.

The **fan-out** (Figure 5.7(b)) of a logic gate is the maximum number of similar gates that may be connected at the output without affecting the performance.

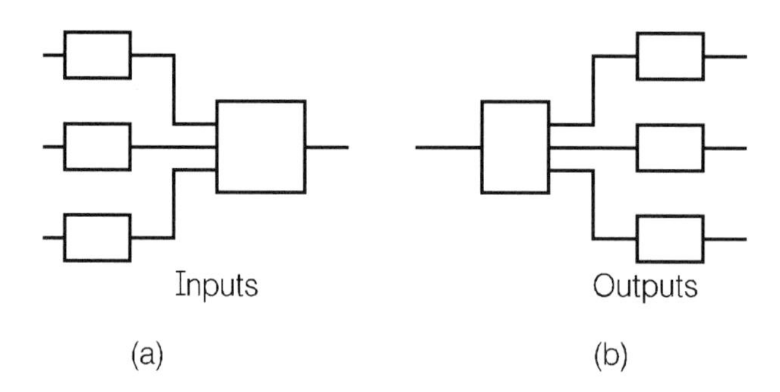

Figure 5.7 *Fan-in and fan-out*

Practical Exercise 5.1

To investigate fan-out
For this exercise you will need the following components and equipment:

6 – silicon diode (1N4148)
3 – resistor (1 kΩ)
1 – +5 V DC power supply
1 – DC voltmeter

Procedure

1. See Figure 5.8 Connect up each circuit in turn and check their operation by copying out and completing the appropriate table.
   Note. For circuit (a) logic 1 will be +5 V and logic 0 approximately +0.7 V, while for circuit (b) logic 1 will be approximately +4.3 V and logic 0 will be 0 V.
2. See Figure 5.9 Check the result of connecting one OR gate to the output of the AND gate (a fan-out of 1) and then the result of adding the second (a fan-out of 2). Work through and complete the table.

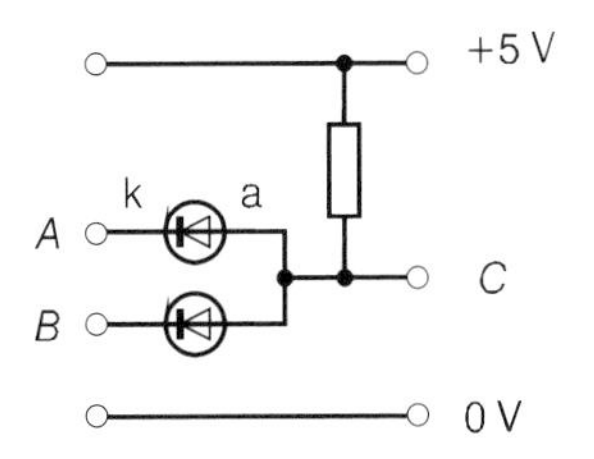

A voltage	A logic level	B voltage	B logic level	C voltage	C logic level
0 V	0	0 V	0		
0 V	0	5 V	1		
5 V	1	0 V	0		
5 V	1	5 V	1		

(a) The 2-input AND gate

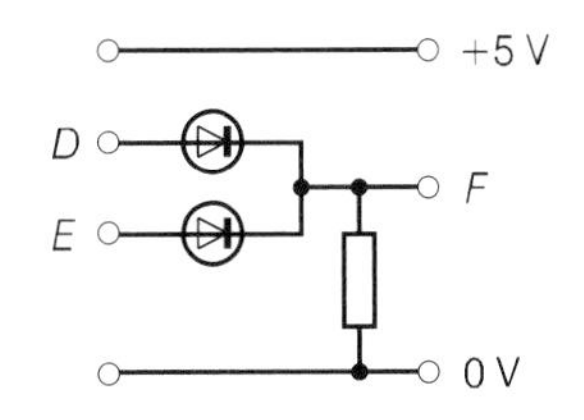

D voltage	D logic level	E voltage	E logic level	F voltage	F logic level
0 V	0	0 V	0		
0 V	0	5 V	1		
5 V	1	0 V	0		
5 V	1	5 V	1		

(b) The 2-input OR gate

Figure 5.8 *Circuit diagrams and tables for Practical Exercise 5.1, procedure 1*

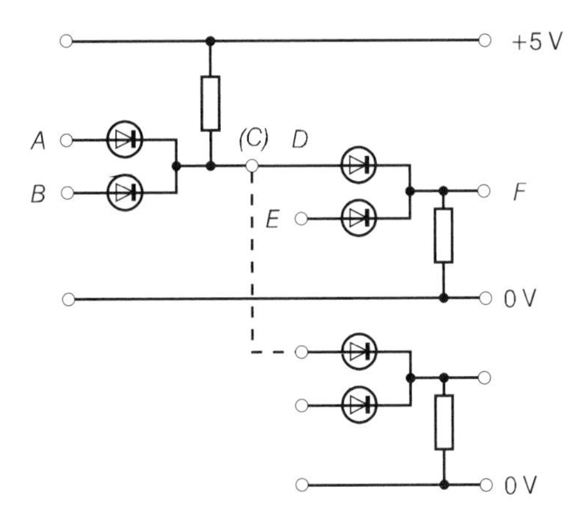

fan out	A voltage	A logic level	B voltage	B logic level	D(=C) voltage	D(=C) logic level	E voltage	E logic level	F should be voltage	F should be logic level	F is voltage	F is logic level
1	0 V	0	0 V	0			0 V	0				
	5 V	1	5 V	1			0 V	0				
2	0 V	0	0 V	0			0 V	0				
	5 V	1	5 V	1			0 V	0				

* Compare these values with the minimum value from the table in figure 5.3

Figure 5.9 *Circuit diagram and table for Practical Exercise 5.1, procedure 2*

Noise margin

Electrical noise is defined as **unwanted signal** and its presence at the input of a digital system can cause false operation of that system. Noise can be produced in the form of a voltage spike which, when superimposed on top of the wanted input signal, causes the output state to change even though the input state has remained constant.

The **noise margin**, also called **noise immunity**, is the maximum noise voltage that can be present at the input without causing any change in the output state.

When two gates are connected (in series) this margin is the difference between the voltage level at the output of the driving gate and the threshold input voltage of the driven gate. See Figure 5.10.

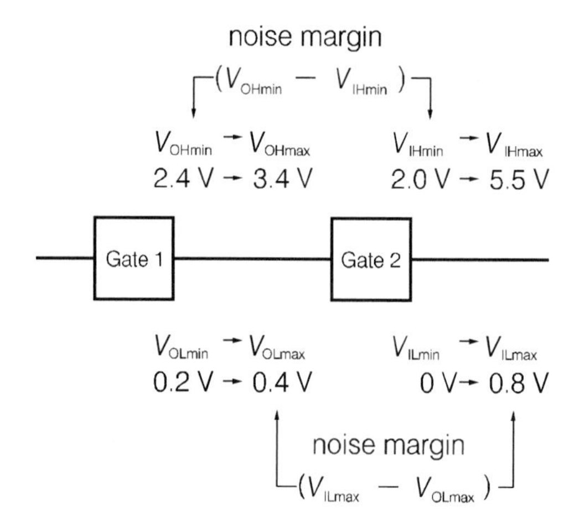

Figure 5.10 *Noise margins*

Two situations are considered in turn.

The logic 1 or high state noise margin (NMH)

The general expression for NMH is given by

Noise margin (NMH) $= V_{OH} - V_{IH}$

This will have a minimum or worst-case value when V_{OH} is a minimum, which, combined with the high state threshold (minimum) value for V_{IH}, results in

$$V_{OHmin} - V_{IHmin} = 2.4\text{ V} - 2.0\text{ V}$$
$$= 0.4\text{ V}$$
$$= 400\text{ mV}$$

The logic 0 or low state noise margin (NML)

The general expression for NML is given by

Noise margin (NML) $= V_{IL} - V_{OL}$

This will give a worst-case figure when V_{OL} is a maximum and, combined with the low state threshold (maximum) for V_{IL}, results in

$$V_{ILmax} - V_{OLmax} = 0.8\ V - 0.4\ V$$
$$= 0.4\ V$$
$$= 400\ mV$$

This is the *guaranteed* noise margin for TTL devices. In practice it is typically 1 V. The connection between noise margins and logic levels is illustrated by Figure 5.11.

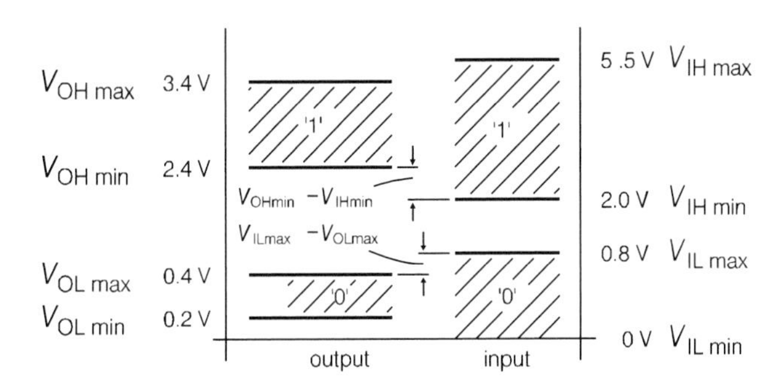

Figure 5.11 *Logic levels and noise margins*

The noise margin for CMOS will depend on the supply voltage (V_{DD}). Remember that CMOS devices can operate with values of V_{DD} from +3 V to +15 V, with logic 0 being from 0 V to 0.3 V_{DD} and logic 1 from 0.7 V_{DD} to V_{DD}. This gives a noise margin of 0.3 V_{DD}.

Practical Exercise 5.2

To investigate logic levels
For this exercise you will need the following components and equipment:

1 – 74LS08 ic (quad 2-input AND gate)
1 – 0 to +5 V (variable) DC power supply
1 – DC voltmeter

Procedure

1 Make up the circuit of Figure 5.12.
The pin connection diagram for the 7408 ic is given in Figure 2.13(a). Remember that both inputs must be logic 1 for the output to be logic 1.

2 Before taking any measurements, investigate the action of the output as the input voltage is increased from 0 V towards +5 V. You should notice that the

Continued on p. 72

Practical Exercise 5.2 (*Continued*)

logic 0 level (V_{OL}) is maintained at the output of the AND gate over a range of input voltages (less than V_{IHmin}) and at a certain input threshold value (V_{IH}) there is a sudden rise in the output (V_{OH}).

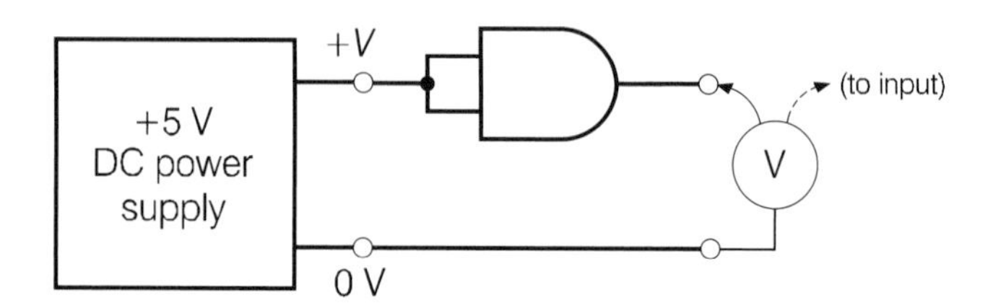

Figure 5.12 *Logic levels: circuit diagram for Practical Exercise 5.2*

3. Increase the input voltage from 0 V towards +5 V and note the value of the input (V_{IH}) at which the output changes (to V_{OH}) and the value of this output. You should expect values of approximately 2 V and 4 V respectively.
4. Calculate the high state noise margin of the AND gate.

Power dissipation

This is the last, but by no means the least, of the factors to be considered when choosing a digital device. Power is dissipated or used in a transistor only when it is actually switching from one state to another.

Important points

- Once switching is complete and the transistor is (say) in the 'off' state, no collector current (I_C) flows and therefore no power ($= V_{CE}I_C$) is used.
- The opposite situation exists when the transistor is fully on, because in this case the collector-emitter voltage (V_{CE}) is zero, and again $V_{CE}I_C = 0$.
- Power is *dissipated* in the circuit resistors.
- The total DC power dissipation per gate (supply voltage × mean supply current) varies from 10 microwatts to 100 milliwatts, depending on the logic family.

Speed–power product

The speed–power product is an interesting characteristic, since it reinforces the fact that a gate with lower power dissipation (CMOS 4000) will have a larger propagation delay

TYPE	PROPAGATION DELAY (ns)	POWER DISSIPATION (mW) (i)	FAN OUT	SUPPLY VOLTAGE (V)	SPEED–POWER PRODUCT (pJ)	NOISE IMMUNITY (V)
Standard TTL (74)	10	10	10	5	100	0.4 (ii)
Low Power Schottky TTL (74LS)	10	2	20	5	20	0.4 (ii)
Advanced low power Schottky TTL (74ALS)	4	1	40	5	4	0.4 (ii)
High speed CMOS TTL (74 HC)	8	0.001	8	5	0.008	$0.3xV_{DD}$
CMOS (4000)	105 (iii)	0.001	50	5	0.105	$0.3xV_{DD}$
ECL (10k)	2	40	30	–5.2	80	0.4 (ii)

Notes: (i) Per gate
(ii) This is the WORST case
(iii) The delay is LESS at HIGHER supply voltages

Figure 5.13 *Characteristics of the various logic families*

(than TTL) – the quantities speed and power being inversely related. If a reminder is needed, the product of ns ($n = 10^{-9}$) and mW ($m = 10^{-3}$) gives pico watt-second (pWs) or picojoules (pJ), where $p = 10^{-12}$.

TTL logic

The basic TTL circuit

This is the most popular logic family, being easily available at low cost and good performance. It uses bipolar technology. The basic TTL circuit is shown in Figure 5.14.

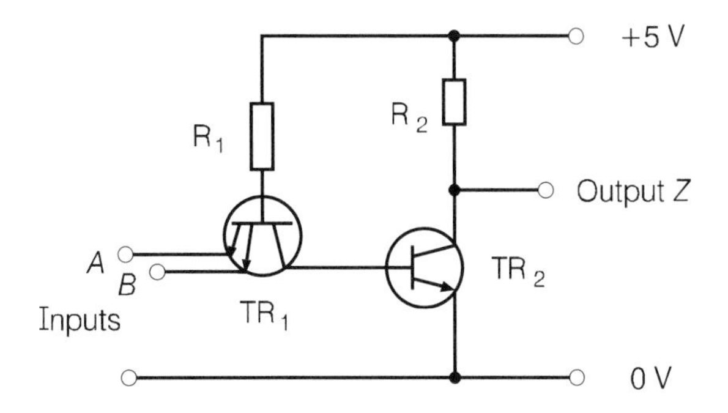

Figure 5.14 *The TTL NAND gate: the basic circuit*

Important points

- TR_1 is a multi-emitter transistor. In this example there are two emitters, which therefore comprise the fan-in of this device.
- The inputs are to the emitters of the transistor TR_1 with the base being connected to a positive voltage V_{CC} through the resistor R_1.
- When the base is more positive than the emitter, the base–emitter junction is **forward biased**.
- When the collector is more positive than the base, the collector–base junction is **reverse biased**.

Circuit operation

(a) With *all* inputs at +5 V (logic 1):

- The emitters must be at the same or greater potential as the base, thus *all* base-emitter junctions will be **reverse biased**.
- The collector voltage of TR_1 (and the base voltage of TR_2) will be high (greater than +0.7 V), the emitter of TR_2 is at 0 V, thus the base–collector junction of TR_1 will be forward biased.
- Current flows into the base of TR_2 through R_1 and the supply.
- Transistor TR_2 will be **on** and its collector voltage (the TTL output) will be low (typically +0.2 V), which means logic 0.

(b) With *any* (or *all*) input(s) at 0 V (logic 0):

- That particular base–emitter junction will be **forward biased**.
- Current will flow in that junction and the collector voltage of TR_1 will be low (less than +0.7 V) and insufficient to forward bias TR_2, which will therefore be **off** and whose collector voltage (the TTL output) will be high (+5 V), that is, logic 1.

(c) Transistor TR_1 acts as an AND gate while transistor TR_2 performs the NOT function, thus giving an overall NAND gate.

The standard TTL NAND gate with totem pole output

See Figure 5.15.

The circuit operation continues:

(a) With *all* inputs at +5 V (logic 1):
(The collector–base junction of TR_1 is forward biased, current flows into the base of transistor TR_2, which will therefore be **on**, having a collector voltage of almost zero.)

- The voltage at base (and emitter) of TR_3 will be almost zero, thus TR_3 will be **off**.

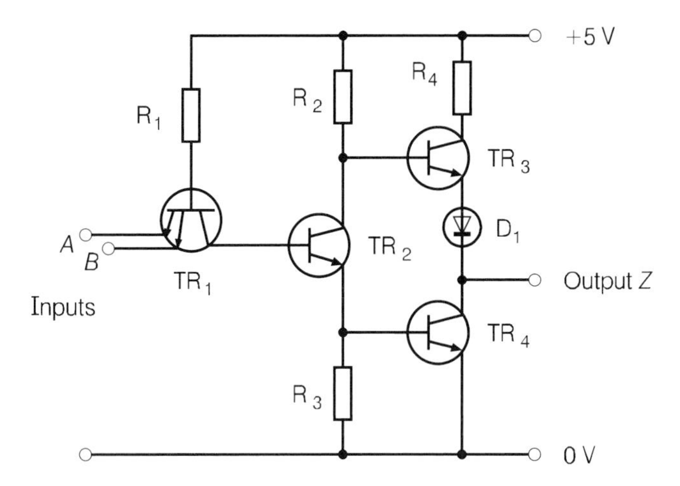

Figure 5.15 *The TTL NAND gate with totem pole output stage*

- The current through TR_2 and R_3 will provide sufficient base current for TR_4 to **turn on**, giving a collector voltage for TR_4 of almost zero and a logic 0 output for Z.
- Diode D will not conduct.

(b) With *any* input at 0 V (logic 0):
(Transistor TR_2 is **off** with a collector voltage of +5 V.)

- The emitter voltage of TR_2 will be zero, TR_4 will be **off** (collector voltage high), giving a logic 1 at the output.
- Meanwhile, transistor TR_3 will turn **on**, its emitter voltage will 'follow' that of its base and diode D will conduct.
- The actual voltage level at the output will be +5 V minus the combined volt-drop across TR_3 and diode D.

Important points

- The additional stage TR_3 and TR_4 is known as a 'totem pole' output. (A dictionary definition of totem is 'a vertical array'.)
- Transistor TR_2 with resistors R_2 and R_3 acts as a 'phase-splitter', giving opposite logic levels at collector and emitter respectively.
- Diode D is necessary to ensure that TR_3 and TR_4 do not conduct at the same time.
- The totem pole arrangement provides a low impedance output for both logic states, thus increasing the fan-out (to 10). In addition it gives faster switching, which means a reduced propagation delay.
- Diodes are sometimes connected across each input to remove (take to the 0 V line) any negative voltages which may inadvertently be applied.

Practical Exercise 5.3

To investigate the voltage levels of a TTL NAND gate
For this exercise you will need the following components and equipment:

5 – BC109 bipolar transistor
4 – resistor (3.9 kΩ, 1.5 kΩ, 1 kΩ and 120 Ω)
1 – silicon diode (1N4148)
1 – +5 V DC power supply
1 – DC voltmeter

Procedure (a)

1 Make up the circuit shown in Figure 5.16

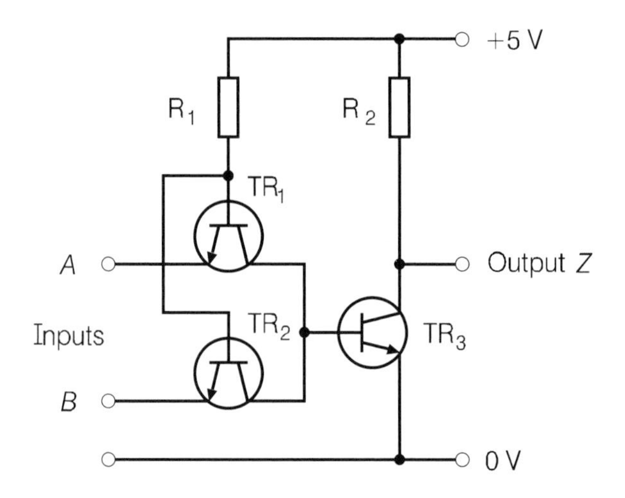

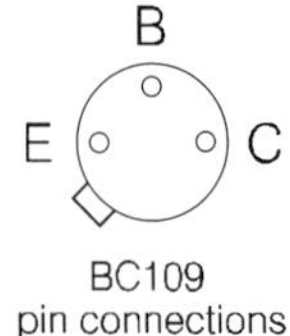

Figure 5.16 *The TTL NAND gate using discrete components: circuit for Practical Exercise 5.3(a)*

2 As a result of taking the appropriate measurements (all with respect to the 0 V line) and doing the necessary calculations, copy out and complete the table shown in Figure 5.17.

A B Inputs V_{E1} V_{E2}	measure $V_{B1} = V_{B2}$	measure $V_{C1} = V_{C2} = V_{B3}$	calculate V_{BE1}	calculate V_{BE2}	calculate $V_{BC1} = V_{BC2}$	measure V_{C3}	state of TR3 (ON/OFF)	Output Z (logic level)
+5 V +5 V (1 1)								
junction is (forward/reverse) biased →								
+5 V +0 V (1 0)								
junction is (forward/reverse) biased →								
+0 V +5 V (0 1)								
junction is (forward/reverse) biased →								
+0 V +0 V (0 0)								
junction is (forward/reverse) biased →								

Figure 5.17 *Results table for Practical Exercise 5.3(a)*

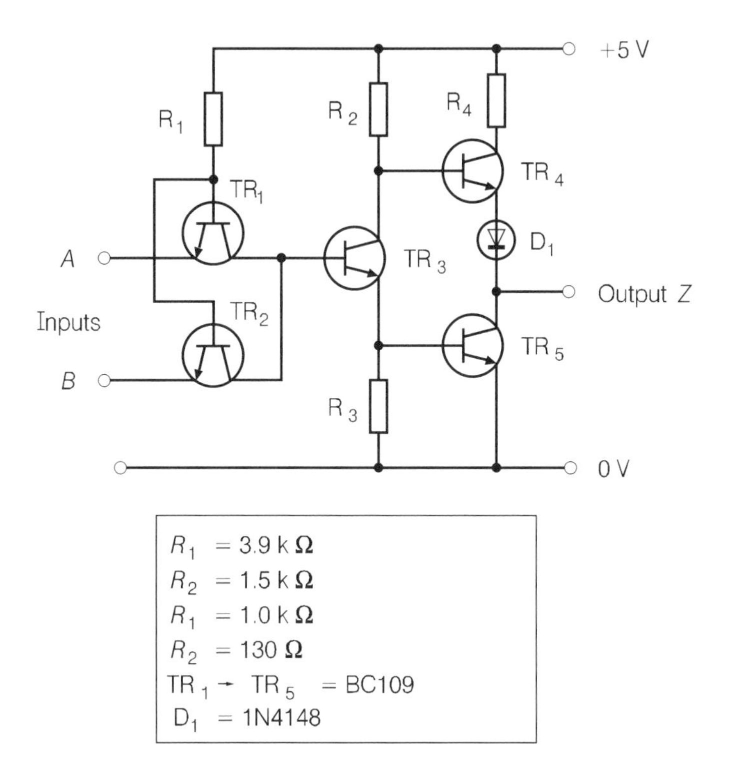

Figure 5.18 *The TTL NAND gate with totem pole output stage using discrete components: circuit for Practical Exercise 5.3(b)*

Continued on p. 78

Practical Exercise 5.3 (*Continued*)

Procedure (b)

1 Make up the circuit shown in Figure 5.18. For the BC109 pin connections see within Figure 5.16.
2 Take measurements, do the calculations, and hence complete the table shown in Figure 5.19.

Inputs *A*	*B*	measure		calc.	state of TR3 (ON/OFF)	measure		calc.	state of TR4 (ON/OFF)	measure	Output *Z* (logic level)
V_{E1}	V_{E2}	V_{B3}	V_{E3}	V_{BE3}		V_{B4}	V_{E4}	V_{BE4}		V_{C4}	
+5 V (1	+5 V 1)										
+5 V (1	+0 V 0)										

Figure 5.19 *Results table for Practical Exercise 5.3(b)*

The open collector gate

It is sometimes necessary to achieve one of the following:

(a) increase the fan-out;
(b) change the logic function;
(c) connect the gates to a common output line or bus.

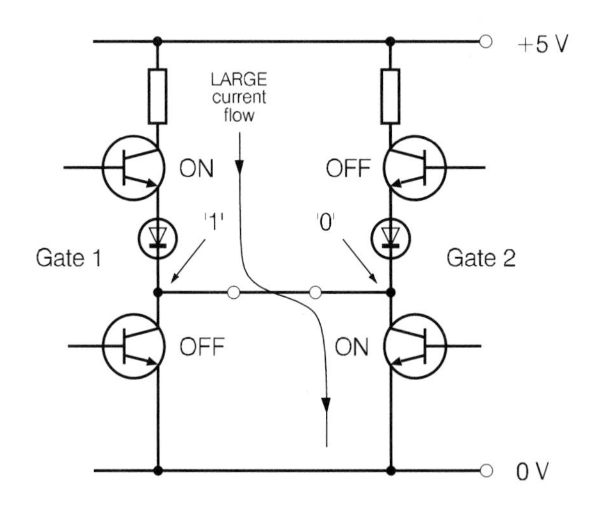

Figure 5.20 *The problems when totem pole outputs are connected together*

It would be reasonable to suggest that, in order to achieve the above objectives, all that is necessary is to connect sufficient TTL totem-pole gates in parallel. Would this work?

The answer is – no, it would not. There would be no problem if the outputs were all at the same logic level, but this is unlikely to be the case. Connecting together two different logic (and hence voltage) levels can cause a large current to flow in the output transistors, with resultant damage to them. To understand why this is so, look at Figure 5.20 in which gates 1 and 2 have high (logic 1) and low (logic 0) outputs respectively. The low resistance path for the current is provided by the two 'on' transistors and, for the situation shown, it is these transistors which are at risk. In particular, the 'on' transistor in gate 2 must accept (act as a sink for) not only the current from the other 'on' transistor but also that from the external load. An excessive current could thus damage either or both of the gates.

The problem can be solved by using TTL gates in which the output stage is replaced by a transistor without a collector load, Figure 5.21. This is known as an **open collector**

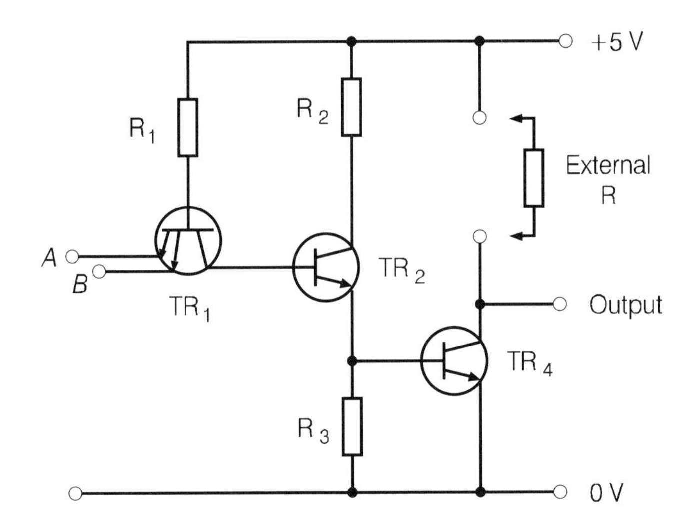

Figure 5.21 *The open collector TTL NAND gate*

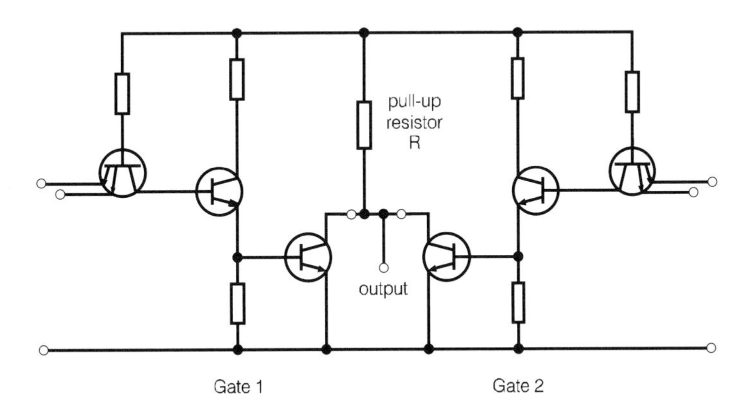

Figure 5.22 *Open collector TTL NAND gates in parallel*

arrangement. A so-called **pull-up resistor** (R), which ensures that the output will go high when the transistor turns off, completes the collector circuit of TR_4 and is provided externally. The value of this resistor will depend on the number of gates connected together. These gates can then be wired in parallel as shown in Figure 5.22. The arrangement is known as *wired logic* and an example now follows.

The wired AND/OR gate

Figure 5.23 shows two open-collector NAND gates wired together. The NAND gates will both provide the NOT (inverter) function, since the inputs of each are connected together. If the output of either gate ($\overline{A}$ or $\overline{B}$) or both gates is at logic 1 (+5 V), the output Z will also be at logic 1. When, however, both outputs are logic 0 the output will be logic 0.

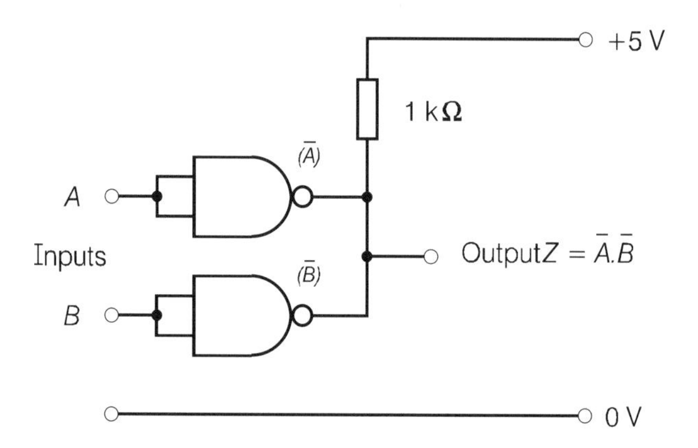

Figure 5.23 *The wired AND/OR gate*

The truth table is shown in Figure 5.24. From the truth table we can deduce that the output expression is given by

$$Z = \overline{A} \cdot \overline{B}$$

which is the AND function with inputs $\overline{A}$ and $\overline{B}$.

Using De Morgan's theorem we can take this further, since

$$Z = \overline{A} \cdot \overline{B} = \overline{A + B}$$

A	B	Z	
0	0	1	← $Z = \overline{A}.\overline{B}$
0	1	0	
1	0	0	
1	1	0	

Figure 5.24 *Truth table for the wired AND gate*

which is the NOR function from the point of view of A and B. But that can be seen from the truth table anyway. Strangely, perhaps, the above gate is sometimes known as a **wired-OR**!

The practical exercise which follows will, among other things, verify the above statements for Z.

Practical Exercise 5.4

To investigate the wired AND gate

For this exercise you will need the following components and equipment:

1 – 74LS03 ic (quad 2-input NAND with open collector)
1 – +5 V DC power supply
1 – LED (5 mm) and resistor (270 Ω)
1 – resistor (1 kΩ)

Procedure (a)

1 Make up the circuit of Figure 5.25. The pin connection diagram for the 7403 ic is identical to that for the 7400 ic given in Figure 2.13(b).

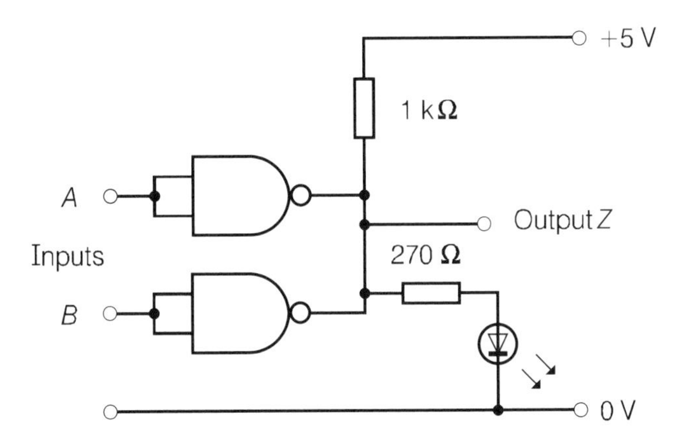

Figure 5.25 *The wired AND gate: circuit for Practical Exercise 5.4(a)*

2 Work through the truth table of Figure 5.24 and hence justify the earlier statements.

Procedure (b)

1 Make up the circuit of Figure 5.26.

Continued on p. 82

Practical Exercise 5.4 *(Continued)*

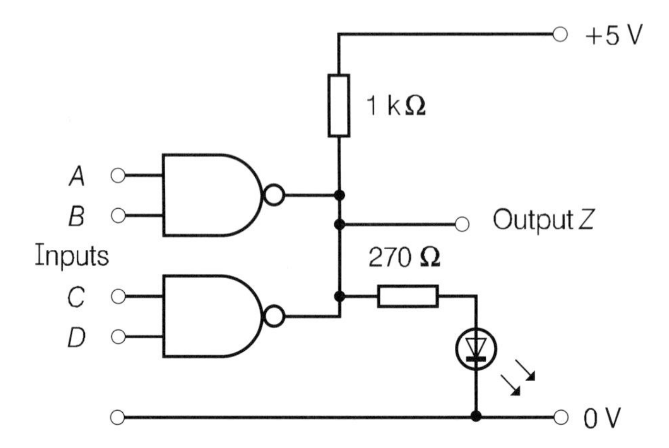

Figure 5.26 *The wired AND gate: circuit for Practical Exercise 5.4(b)*

2 Investigate the action by drawing up and working through the truth table for the four input variables *A*, *B*, *C* and *D*. Complete the column for output *Z* and hence show that this gate will provide the AND function for inputs $\overline{A \cdot B}$ and $\overline{C \cdot D}$ given by

$$Z = \overline{A \cdot B} \cdot \overline{C \cdot D}$$

Hint. It would be helpful if the truth table contained the columns for

$\overline{A \cdot B}$ $\overline{C \cdot D}$ $\overline{A \cdot B} \cdot \overline{C \cdot D}$ and $\overline{A \cdot B + C \cdot D}$

3 Use a De Morgan theorem on the expression for *Z* in (2) above and, as a result, show that the OR (NOR) equivalent is given by

$$Z = \overline{A \cdot B + C \cdot D}$$

Sourcing and sinking

The situation exists where the output of one TTL (driving) gate is connected to the input of another TTL (driven) gate, as in Figure 5.27(a). The direction of the current flow will depend upon the voltage level at the output of the driving gate.

Suppose (Figure 5.27(a)) that the output of the driving gate is high (positive voltage = logic 1). Current will then flow from the output of the driving gate to the input of the driven gate. In other words, the driving gate supplies current to the driven gate. This is known as a **sourcing current**.

If, on the other hand (Figure 5.27 (b)) the output of the driving gate is low (zero voltage = logic 0). then current will flow from the input of the driven gate into the output of the driving gate. In this case the driving gate takes current from the driven gate. This is known as a **sinking current**.

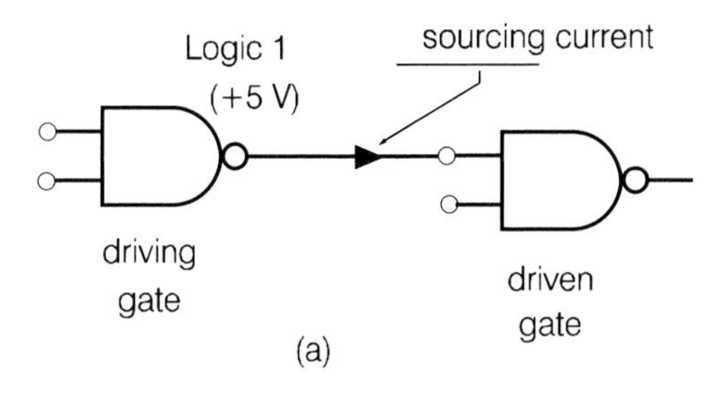

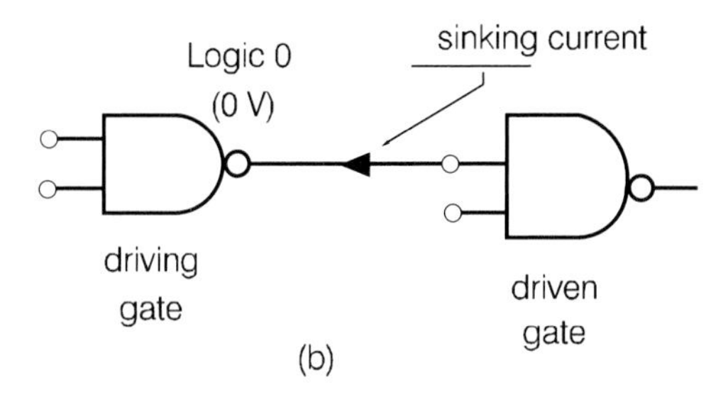

Figure 5.27 *Sourcing and sinking*

Important points

- As a convention, current entering a gate terminal is regarded as positive, whereas current leaving a gate terminal is regarded as negative.
- Manufacturers' data gives positive values for sinking currents and negative values for sourcing currents.

Developing the idea further we come to Figure 5.28 and take the standard TTL gate as the example. Reference to manufacturers' data (e.g. Maplin) gives the following (typical) information.

Sourcing (output of driving gate is high, input of driven gate is high)

- Maximum current supplied (sourced) by driving gate (I_{OH}) will be -400 μA. (Note the minus sign for current leaving a gate.)
- Current required by driven gate (I_{IH}) is 40 μA.
- The maximum number of inputs (gates) that can be driven, i.e. the fan-out of the driving gate in the high state, is given by

$$\text{fan-out} = I_{OH}/I_{IH} = 400\ \mu\text{A}/40\ \mu\text{A} = 10$$

Sinking (output of driving gate is low, input of driven gate is low)

- Maximum current taken by the driving gate (I_{OL}) is 16 mA. (The gate can 'sink' 16 mA.)

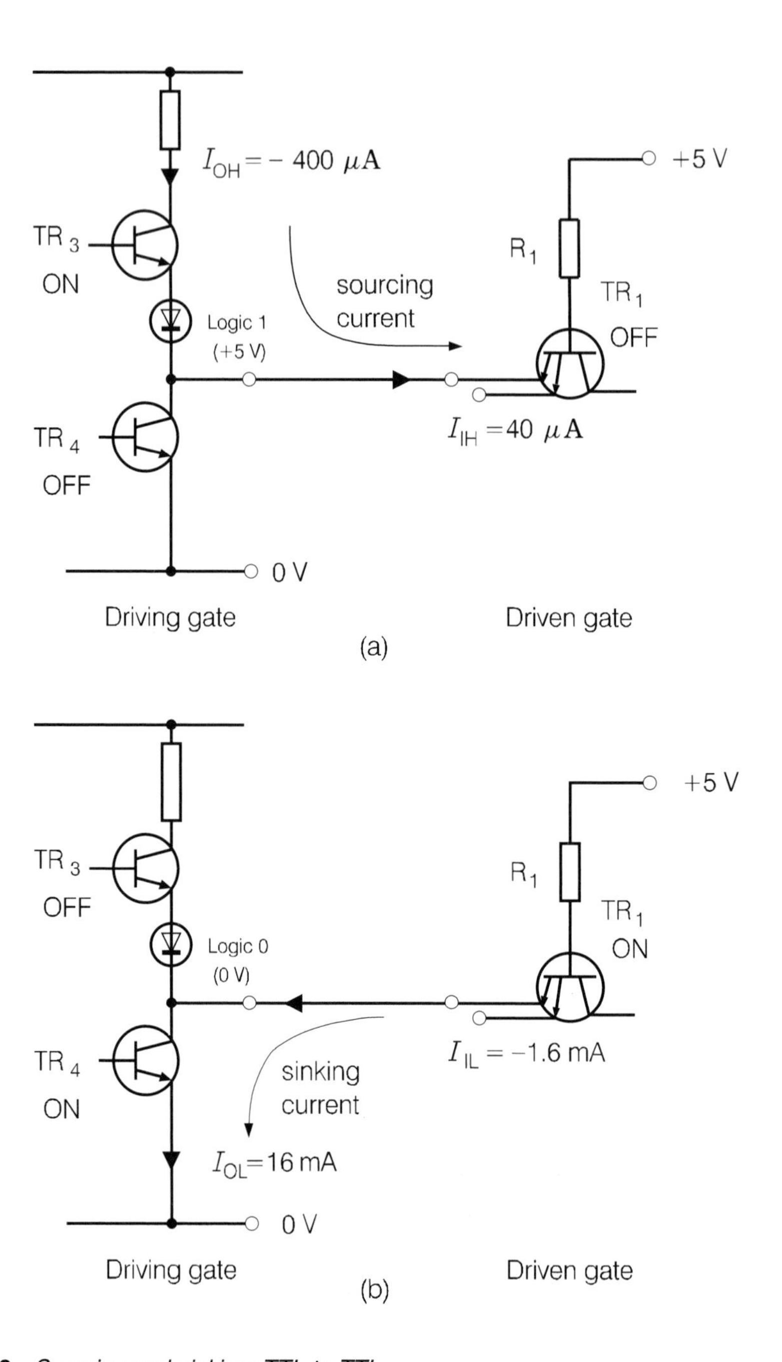

Figure 5.28 *Sourcing and sinking: TTL to TTL*

- The input current of the driven gate (I_{IL}) is −1.6 mA. (Note again the minus sign, referring to current leaving a gate.)
- The fan-out of the driving gate in the low state is given by

$$\text{fan-out} = I_{OL}/I_{IL} = 16\text{ mA}/1.6\text{ mA} = 10$$

With the data used, both states give the same fan-out value. In cases where this is not so, the lower value for fan-out should be taken in order not to overload the driving gate. The result of overloading is to cause *either* the output (high level) to fall below its minimum value of 2 V (remember Practical Exercise 5.2) *or* the output (low level) to rise above its maximum value of 0.8 V.

The pull-up resistor

The **minimum value** of the pull-up resistor R is determined from the logic 0 output state and the required fan-out, since the total current that can be taken by the driving gate (I_{OL}) is 16 mA.

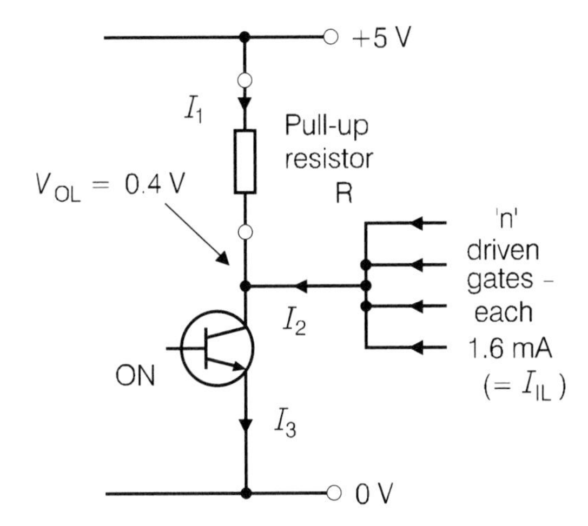

Figure 5.29 *The minimum value of the pull-up resistor*

Figure 5.29 shows that:

$$R_{min} = (V_{CC} - V_{OL})/I_1$$
$$= (V_{CC} - V_{OL})/(I_3 - I_2)$$

If the required fan-out is, say, 6 ($= n$), then

$$I_2 = 6 \times 1.6 \text{ mA}$$
$$= 9.6 \text{ mA}$$

The data sheet (Figure 5.3) gives $V_{OL} = 0.4$ V.

$$I_3 = I_{OL} = 16 \text{ mA}$$

Hence

$$R = (5 - 0.4) \text{ V}/(16 - 9.6) \text{ mA}$$
$$= 4.6 \text{ V}/6.4 \text{ mA}$$
$$= 719\ \Omega\ (680\ \Omega \text{ nearest preferred value})$$

To calculate the **maximum value** of R it is necessary to consider the logic 1 state of the output.

Suppose that just two gates are connected together (Figure 5.30). Both transistors are off and the current through each of them will be just the leakage current (I_3), given as (typically) 250 μA. The current through $R(I_1)$ will be the sum of the leakage currents ($2I_3$) and the total current into the driven gates (I_2).

$$R_{max} = (V_{CC} - V_{OH})/I_1$$

$$= (V_{CC} - V_{OH})/(I_2 + 2I_3)$$

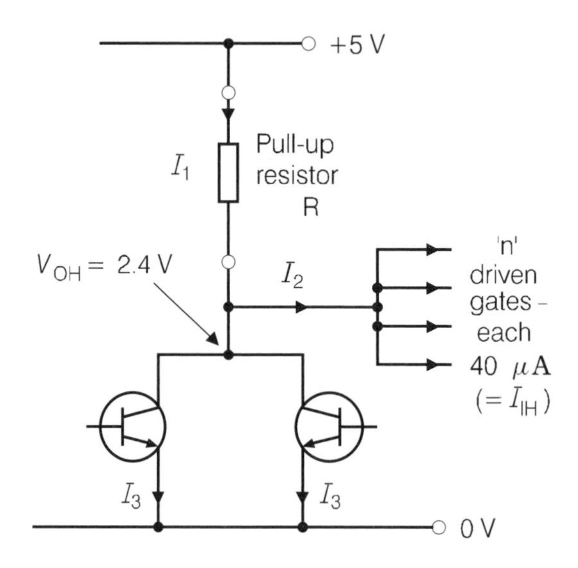

Figure 5.30 *The maximum value of the pull-up resistor*

If the required fan-out is still 6, then

$$I_2 = 6 \times 40\ \mu A$$

$$= 240\ \mu A$$

$$= 0.24\ mA$$

$$2I_3 = 500\ \mu A$$

$$= 0.5\ mA$$

The data sheet gives $V_{OH} = 2.4$ V. Hence

$$R_{max} = (5 - 2.4)\ V/(0.24 + 0.5)\ mA$$

$$= 2.6\ V/0.74\ mA$$

$$= 3514\ \Omega\ (3300\ \Omega \text{ nearest preferred value})$$

These calculations show that the value for the pull-up resistor can be between 680 Ω and 3300 Ω.

Questions

Both questions assume a supply voltage of +5 V. The manufacturers' data given earlier should be used to provide other basic information.

5.1 Calculate the maximum and minimum value for the pull-up resistor for standard TTL open-collector gates, to provide a fan-out of 8, if the number of connected gates is (a) 2, (b) 4.

5.2 Three standard open-collector TTL NAND gates are wired together and the fan-out is 2. Calculate the range of values for the pull-up resistor.

Tristate logic devices

Important points

Remember that:

- The output of a normal logic gate will be either logic 0 or logic 1 and as a result will either **sink** current from, or **source** current to, a load.
- The totem pole outputs of two TTL gates **cannot** be connected together due to the possibility of excessive power dissipation. The use of the open collector with pull-up resistor helps to overcome this restriction, but the maximum operating speed of the system is reduced.

Tristate gates have the same function as normal gates, with the addition that the output can be switched into a **third** (tri) high impedance state, thus being in neither a sinking nor a sourcing mode. This state is achieved by means of a control or **enabling** input. With the enable input at logic 0, there is normal logic gate operation. Having the enable input at logic 1 causes the output impedance of the gate to be high, with an undefined logic level. The gate output is then, in effect, open circuit or **disabled**.

The result is that a number of tristate devices can be connected to a common line with only one of them being in operation, or enabled, at any given time. This arrangement finds application in computer systems where a large number of devices may be connected to a bidirectional 'bus' line, as shown in Figure 5.31.

A typical device in the TTL range is 74LS541, with the CMOS version being 74HCT541. These contain eight buffers (inverters), all with 3-state outputs, which can be interfaced directly with a system bus.

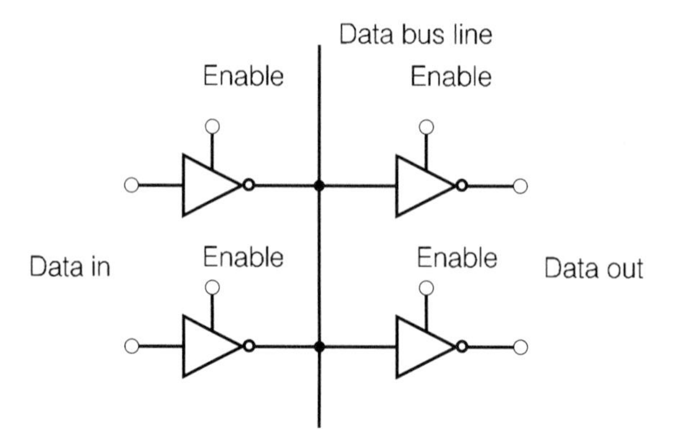

Figure 5.31 *Tristate logic*

Important points

For the standard TTL output:

- The sinking current capacity (16 mA at logic 0) is greater than that for sourcing (400 μA at logic 1).
- For external loads requiring a current greater than 16 mA, some form of interface is required between TTL output and load. A typical arrangement, shown in Figure 5.32, would be a transistor driver working as a current amplifier.

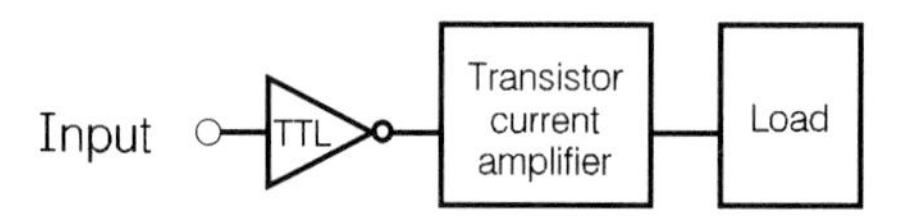

Figure 5.32 *TTL interface to load*

CMOS Logic

This is the 4000 series, possessing low power dissipation, good (high voltage) noise immunity, large fan-out. The downside is that the switching speed is (or was) limited because of the very high input impedance of the device and its input capacitance. High speed versions (suffix HC and AC) are now available.

The manufacture is based on MOS (metal oxide semiconductor) field-effect transistors (MOSFETs), which are unipolar devices. MOS transistors use either p-channel (earlier type) or n-channel material, the latter, because of its faster switching speed, being preferred. Both types possess high input impedance (typically 10^{10} Ω).

nMOS (n-channel) needs a positive supply (hence is compatible with TTL) and will conduct when the gate is positive with respect to the source. Compare this with the forward biased base–emitter junction of the npn bipolar transistor.

pMOS (p-channel), like the pnp transistor, needs a negative supply, conducting when its gate is negative with respect to its source.

The complementary type (CMOS) combines pMOS and nMOS devices. They are cheap to make, consume less power than TTL but generally have a longer switching time and higher noise margin. The small fabrication area required means that they can be made using LSI technology and they find application, for example, as computer memories.

The CMOS NOT gate (inverter)

See Figure 5.33.

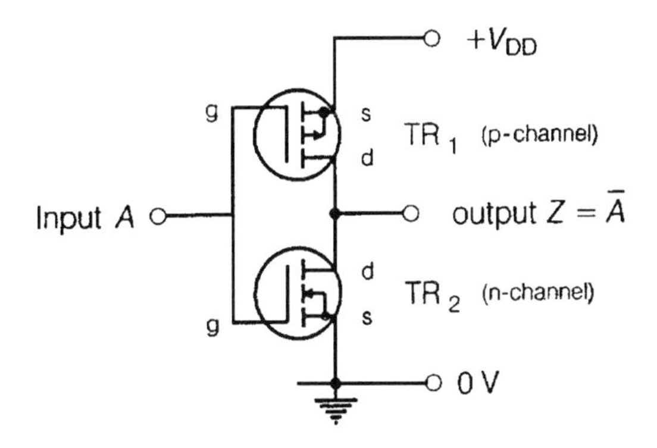

Figure 5.33 *The CMOS NOT gate*

TR_1 is a p-channel device conducting when the gate (g) is negative with respect to the source (s).

TR_2 is an n-channel device conducting when the gate is positive with respect to the source.

When input A is zero (logic 0), TR_2 will be 'off', TR_1 will be 'on' and the output Z will be at approximately $+V_{DD}$ (logic 1).

When input A is positive (logic $1 = +V_{DD}$), TR_1 will be 'off', TR_2 will be 'on' and output Z will be at zero voltage (logic 0).

For both output logic states the output impedance is low because the ON transistor is saturated.

The action of the CMOS NOT gate is summarized in Figure 5.34.

The CMOS NOR gate

See Figure 5.35

When both inputs A and B are zero (logic 0), TR_3 and TR_4 will be 'on' since their gate–source voltages (V_{gs}) are both negative. The output Z will be at $+V_{DD}$ (logic 1).

When input A is positive (logic 1) and input B zero (logic 0), TR_1 ($+V_{gs}$) and $TR_3(-V_{gs})$ will be 'on'. The output Z will be almost zero (logic 0).

Input A		(p-channel) TR_1		(n-channel) TR_2		Output Z	
voltage	logic level	V_{gs}	state	V_{gs}	state	voltage	logic level
0	0	neg.	ON	0	OFF	$+V_{DD}$	1
$+V_{DD}$	1	0	OFF	pos.	ON	zero	0

n-channel conducts when gate is POSITIVE with respect to source

p-channel conducts when gate is NEGATIVE with respect to source

Figure 5.34 *Summary of action of the CMOS NOT gate*

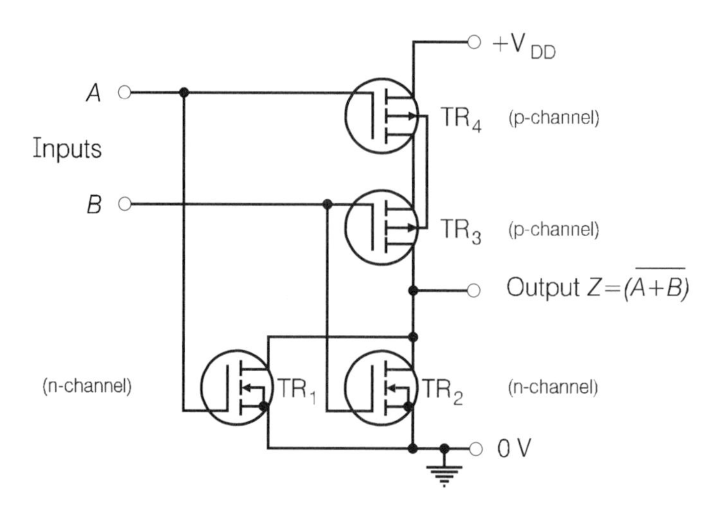

Figure 5.35 *The CMOS NOR gate*

Input B		Input A		(n-channel) TR_1		(n-channel) TR_2		(p-channel) TR_3		(p-channel) TR_4		Output Z	
voltage	logic level	voltage	logic level	V_{gs}	state	V_{gs}	state	V_{gs}	state	V_{gs}	state	voltage	logic level
0	0	0	0	0	OFF	0	OFF	neg.	ON	neg.	ON	$+V_{DD}$	1
0	0	$+V_{DD}$	1	pos.	ON	0	OFF	neg.	ON	0	OFF	zero	0
$+V_{DD}$	1	0	0	0	OFF	pos.	ON	0	OFF	neg.	ON	zero	0
$+V_{DD}$	1	$+V_{DD}$	1	pos.	ON	pos.	ON	0	OFF	0	OFF	zero	0

n-channel conducts when gate is POSITIVE with respect to source

p-channel conducts when gate is NEGATIVE with respect to source

Figure 5.36 *Summary of action of the CMOS NOR gate*

Similarly, with input A zero and input B positive, output Z will again be zero, this time with TR_2 and TR_4 being 'on'.

Finally, with both inputs A and B positive, TR_1 and TR_2 will be 'on' ($+V_{gs}$), with output Z at zero.

The action of the CMOS NOR gate is summarized in the table in Figure 5.36.

The CMOS NAND gate

The CMOS NAND gate (Figure 5.37) can be explained in a similar fashion and it is left for the reader to do this!

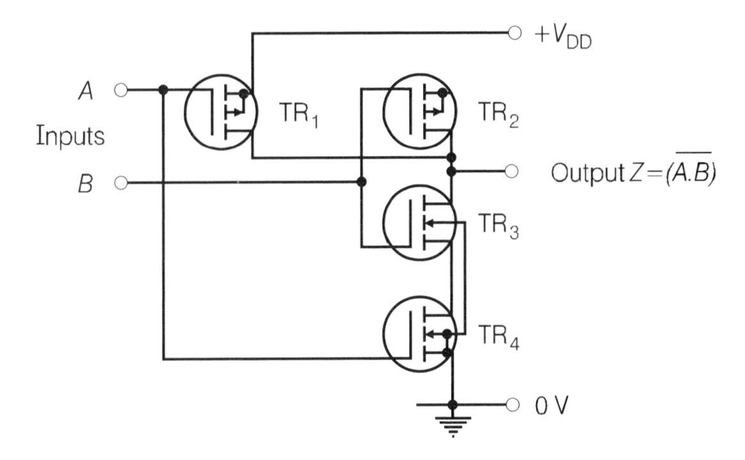

Figure 5.37 *The CMOS NAND gate*

Question

5.3 Explain the action of the CMOS NAND gate, giving your answer in the form of a table similar to that of Figure 5.36.

The ECL OR/NOR gate

This is a non-saturating bipolar family, giving high speed (t_{pD} is typically 2 ns), which makes it the fastest of all logic ic's. The operation is based on the difference or 'long-tailed pair' amplifier.

ECL logic is available in a '10 000' and a '100 000' series, requiring the (unusual) supply voltages of −5.2 V and −4.2 V respectively.

Positive logic is used, with logic 0 = −1.6 V and logic 1 = −0.8 V. Remember that, for positive logic, logic 1 must be **more positive** than logic 0!

The circuit of an ECL OR/NOR gate is given in Figure 5.38.

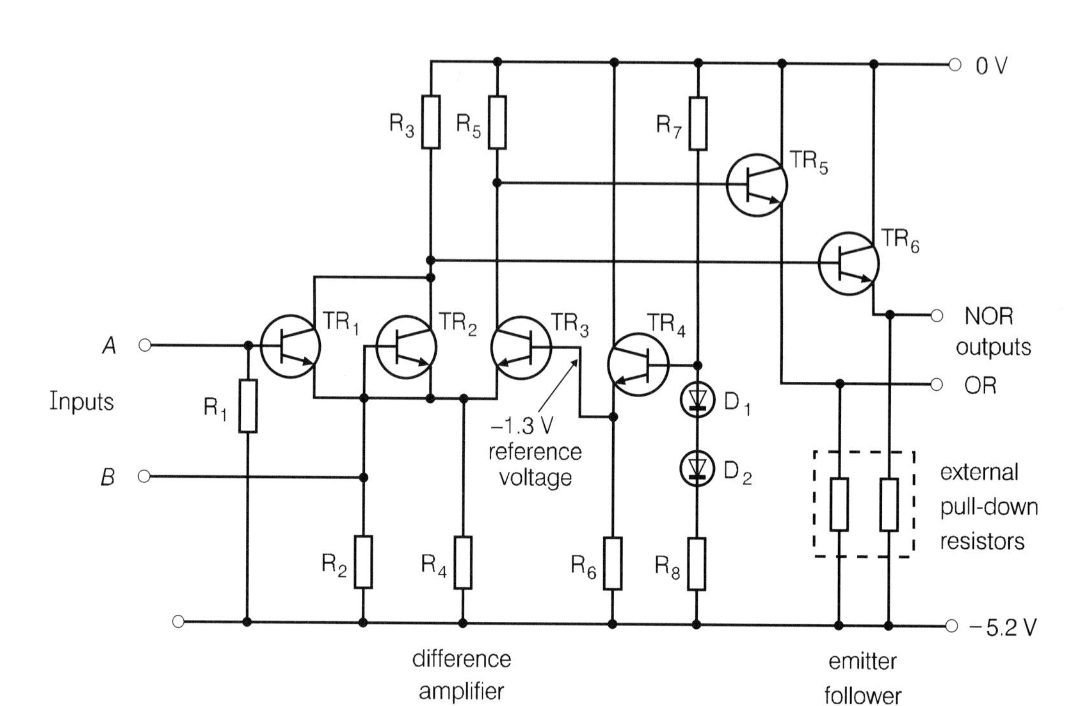

Figure 5.38 *The ECL OR/NOR gate*

The difference amplifier is provided by TR_1, TR_2, TR_3, the common emitter resistor R_4 and the collector loads R_3 and R_5. A reference voltage of -1.3 V is provided at the base of TR_3, made available by the potential divider chain consisting of D_1, D_2, R_7, and R_8 together with transistor TR_4. This will give an emitter voltage for TR_1, TR_2 and TR_3 of $-(1.3 + 0.7)$ V, ie. -2.0 V.

The action is described as follows.

Inputs *A* and *B* at logic 0 (−1.6 V)

V_{BE1} and V_{BE2} are both less than $+0.7$ V, therefore TR_1 and TR_2 are both 'off', giving

$$V_{C3} = 0 \text{ V} = V_{B6}$$

therefore

$$V_{E6} \text{ (NOR output)} = (0 - 0.7) \text{ V} = -0.7 \text{ V} = \text{logic } 1$$

Meanwhile, V_{BE3} is greater than $+0.7$ V, therefore TR_3 is 'on', giving

$$V_{C3} = -1.0 \text{ V} = V_{B5} \text{ (assuming a suitable value of } R_5\text{)}$$

therefore

V_{E5} (OR output) $= -(1.0 + 0.7)$ V $= -1.7$ V $=$ logic 0

One input (say, *A*) at logic 1 (−0.8 V) and the other (*B*) at logic 0 (−1.6 V)

V_{BE} is greater than +0.7 V therefore TR_1 is 'on', giving

$V_{C1} = -1.0$ V $= V_{B6}$ (with suitable R_3)

therefore

V_{E6} (NOR output) $= -(1.0 + 0.7)$ V $= -1.7$ V $=$ logic 0

Meanwhile, V_{BE3} is less than +0.7 V therefore TR_3 is 'off', giving

$V_{C3} = 0$ V $= V_{B5}$

therefore

V_{E5} (OR output) $= (0 - 0.7)$ V $= -0.7$ V $=$ logic 1

The remaining line in the OR/NOR truth table, ($A = B = 1$) can be explained in similar manner.

The **noise margin** for ECL logic is no better than for TTL, being 400 mV, but the fan-out is of the order of 50, due largely to the high impedance of the differential amplifier and the low impedance of the emitter follower output.

Interfacing

Since TTL and CMOS operate on different supply voltages, they cannot normally be used together without consideration being given to the method of interconnection.

TTL circuits need a nominal +5 V supply, regulated to within ±0.25 V for a current of some mA. CMOS, on the other hand, can operate successfully on supply voltages from +3 V to +15 V, with no need for voltage regulation since the supply current is of the order of μA.

Interfacing CMOS to TTL

See Figure 5.39.

When the CMOS gate output is **logic 1**, it will need to supply (source) a current of 40 μA ($= I_{IH}$) to each TTL input gate: no problem for CMOS.

When the CMOS gate output is **logic 0** it will need to accept (sink) a current of 1.6 mA ($= I_{IL}$) from each TTL input gate, without its (CMOS) output rising beyond the permitted maximum (V_{OLmax}) of 0.4 V. Thus the maximum number of TTL gates can be determined. If a direct connection between CMOS and TTL is not possible, then a CMOS buffer is used: better still if this buffer is incorporated in the CMOS chip, e.g. in 4011B the CMOS gate is buffered and can deliver a larger current.

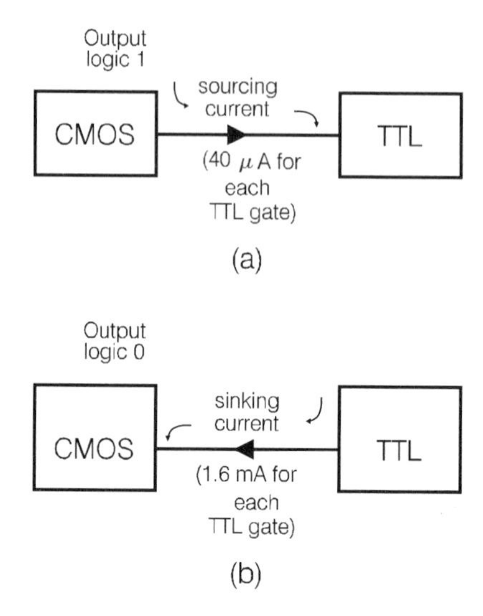

Figure 5.39 *Interfacing: CMOS to TTL*

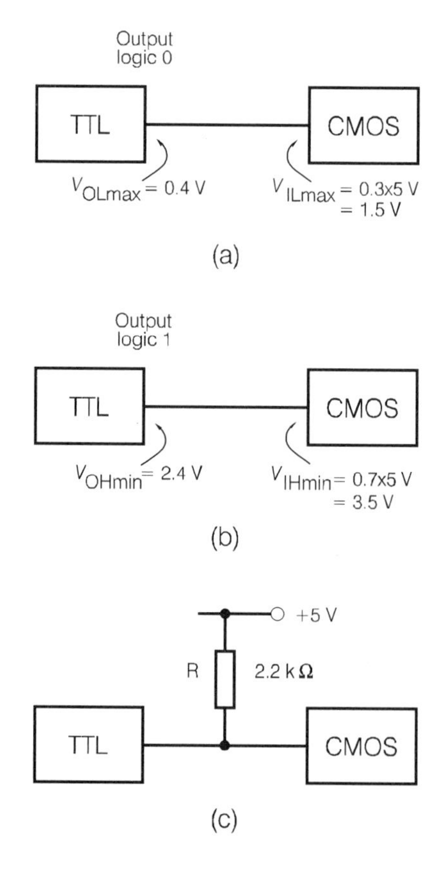

Figure 5.40 *Interfacing: TTL to CMOS*

Interfacing TTL to CMOS

See Figure 5.40.

For a TTL output of logic 0, the low-level threshold (V_{IL}) for CMOS is 0.3 times the supply voltage, i.e. 0.3 × 5 V = 1.5 V. For TTL, $V_{OLmax} = 0.4$ V, and there is no problem since CMOS has very high input resistance and its current demand is negligible.

For a TTL output of logic 1, $V_{OHmin} = 2.4$ V for TTL and $V_{IHmin} = 0.7$ times the supply voltage for CMOS, which amounts to 3.5 V. This is unsatisfactory since it is unlikely that 2.4 V will be accepted by CMOS as logic 1. A pull-up resistor is required, with an open-collector TTL, to raise V_{OH} to about 5 V.

The Schottky diode

This is a device in which the junction is formed between a pure metal such as aluminium and an n-type semiconductor. It has a threshold (turn-on) voltage as low as 0.3 V, charge storage is virtually eliminated and the switching action is very fast. The symbol for the Schottky diode is shown in Figure 5.41(a).

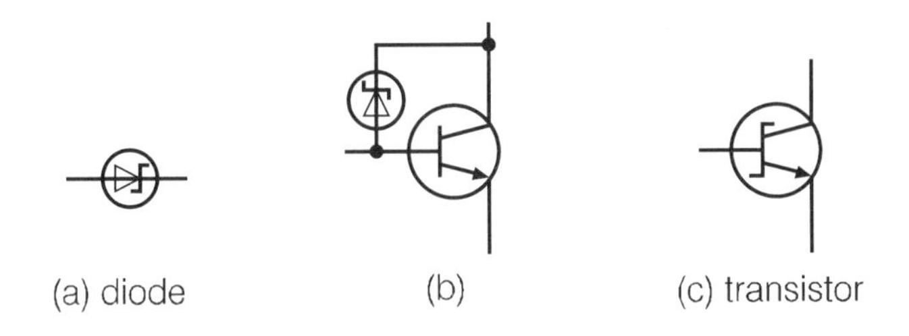

Figure 5.41 *The Schottky diode and transistor: circuit symbols*

Schottky TTL

With standard TTL the transistors saturate, causing a limit on the switching speed which is typically 6 ns ('on' time) and 10 ns ('off' time). In order to decrease this time, the transistors must be prevented from saturating and to achieve this a Schottky diode is connected between base and collector, Figure 5.41(b).

With a positive voltage on the base, TR is turned 'on' (V_{CE} goes to 0.2 V). At this moment the base is more positive than the collector, hence the collector base junction is forward biased. The diode will conduct and divert excess base current away from the base. The transistor receives insufficient base current and is therefore prevented from saturating. This diode prevents excess current entering the base region and hence the transistor does not saturate.

The symbol for the Schottky transistor is shown in Figure 5.41 (c).

The earlier Schottky transistor (74S) is now less popular than the lower power version (74LS) – as used in the practical exercises in this book – and certainly than the advanced low power Schottky (74ALS).

Questions

5.4 Give the basic reason for the relatively slow switching speed of TTL gates and explain how Schottky transistors overcome this limitation of TTL.

5.5 The outputs of standard TTL gates should not be connected together. Why is this? What arrangement must apply before they can be paralleled?

6 Sequential systems 1: Flipflops

The term **sequential system** in logic language means a system in which the resulting output condition is dependent not only on the input but also on the existing output condition. As we shall see later on, the truth tables can be developed from a knowledge of the output state *before* the application of the inputs.

The bistable condition

A flipflop is a device that has two stable conditions or states and which can remain in one or other of these states indefinitely. The device can be made to change state when a trigger signal (pulse) is applied. The application of a second trigger pulse will cause the device to revert to its original state. This two-state situation, known as a bistable state, forms the basis of operation of counters and registers.

Practical Exercise 6.1 allows you to become familiar with the basic bistable operation.

Practical Exercise 6.1

The two-transistor switch
For this exercise you will need the following components and equipment:

2 – BC109 npn transistor
2 – LED (5 mm)
2 – resistor (270 Ω)
2 – resistor (10 kΩ)
1 – 5 V DC power supply
1 – DC voltmeter

Procedure

1 Connect up the circuit shown in Figure 6.1. The connection diagram for the BC109 transistor is given in Figure 5.16.

Continued on p. 98

Practical Exercise 6.1 (*Continued*)

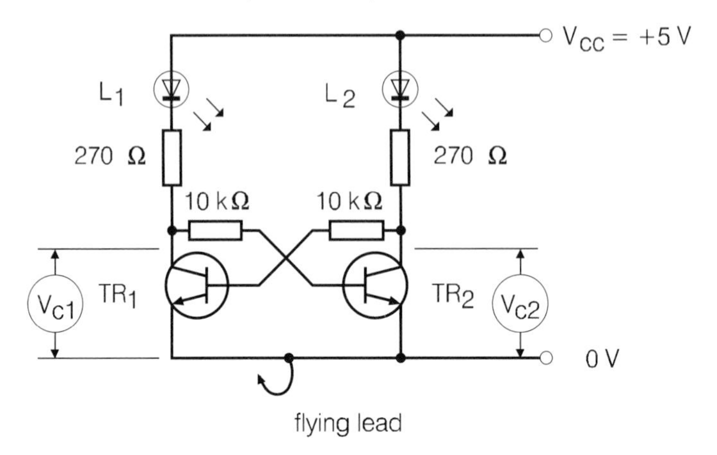

Figure 6.1 *The two-transistor switch: circuit for Practical Exercise 6.1*

2 For the moment, leave the free end of the flying lead disconnected.
3 Switch on the supply and note which one of the two LEDs lights up.
4 Switch the supply off and then on. Again, note which LED lights up. (It should be the same LED on both occasions. The actual LED that lights will depend upon the characteristics of the transistors, which are unlikely to be identical, thus causing an unbalance in the circuit).

The LED that is lit (say, L_1) indicates that the associated transistor (TR_1) is 'on', or conducting, and that the other transistor (TR_2) is 'off', or not conducting.

Action	State of lamps (lit/not lit)		State of transistors (on/off)		Voltage measurement	
	L_1	L_2	TR_1	TR_2	V_{C1}	V_{C2}
Switch on supply						
Flying lead to base of 'on' transistor						
Flying lead to base of 'on' transistor						

Figure 6.2 *Results table for Practical Exercise 6.1*

5 Switch off the 'on' transistor (and hence the 'on' LED) by connecting the end of the flying lead to the base of that transistor. This will zero-bias the base–emitter junction of that transistor and should cause the other transistor to switch on, shown by the associated LED being lit.
6 Repeat the exercise, now taking voltage measurements V_{c1} and V_{c2}. Copy out and complete the table in Figure 6.2.

Important points

- You will notice that the cycle repeats itself, and that after two changes the original state is obtained.
- A series of trigger input pulses causing the transistors to alternately switch on and off would produce a so-called **train** of output pulses, of height $+V_{cc}$ volts.
- A V_{cc} value of +5 V would provide a suitable logic 1 level for TTL.
- The two-state or bistable device can be likened to a see-saw, whose ends must always be going in opposite directions. At the extreme of their travel, one end is on the ground while the other end is up in the air. Both ends can never, at the same time, be either on the ground or up at the highest point!

The latch

The familiar conventional switch is the on/off type, where a simple state of affairs occurs, see Figure 6.3. The lamp will light when the switch is closed, and will remain lit only for as long as the switch remains closed.

Looking now at Figure 6.4, the behaviour of this particular unit allows the following statements to be made:

(a) With both switches (S and R) open, the lamp **will not be** lit.
(b) When the SET switch (S) is closed, the lamp **will be** lit.
(c) The lamp will remain lit when this switch is opened until the RESET switch (R) is closed. The lamp is said to be **latched on**.

In logic terms, we can write (a) to (c) above like this:

(a) With $S = 0$ (open), $R = 0$ (open),

previous lamp state (Q) was $Q = 0$ (off)

present lamp state (Q^+) is $Q^+ = 0$ (off)

(b) With $S = 1$ (closed), $R = 0$ (open),

$Q^+ = 1$ (the SET state)

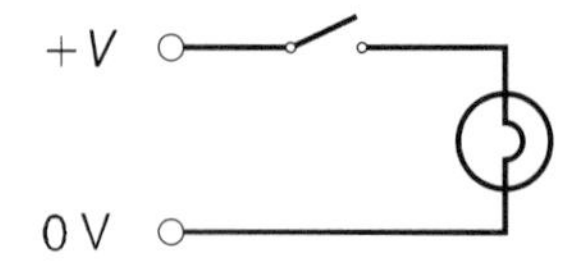

Figure 6.3 *The on–off switch*

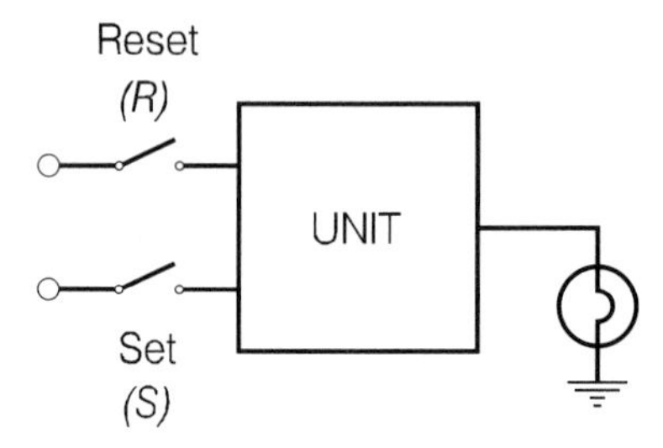

Figure 6.4 *The latch*

(c) With $S = 0$, $R = 0$, $Q^{+} = 1$
With $S = 0$, $R = 1$, $Q^{+} = 0$ (the RESET state)

The truth (state) table is given in Figure 6.5.

R	S	Q	Q^{+}	Comment
0	1	0	1	SET
0	0	1	1	NO CHANGE (Hold)
1	0	1	0	RESET
0	0	0	0	NO CHANGE (Hold)

Figure 6.5 *Truth table for the latch*

Important points

- Q is the output condition before the inputs are applied.
- Q^{+} is the output condition (of Q) immediately after the inputs are applied.

Bistable (or flipflop) circuits

A glance at the catalogues will reveal a variety of available types of bistable devices, such as RS, D, JK and JK master–slave. It is appropriate to begin with the simplest, the RS (or SR) type.

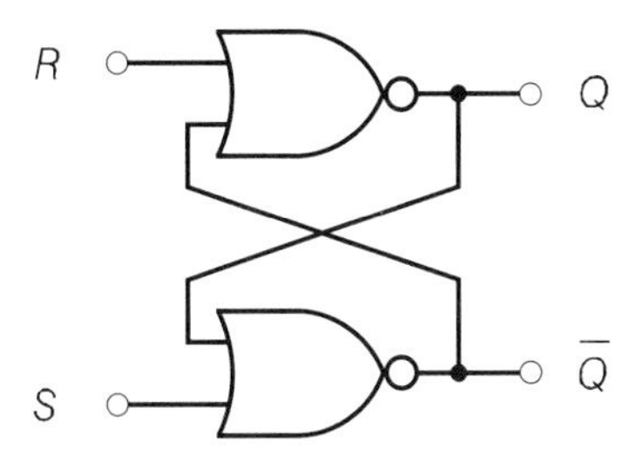

Figure 6.6 *The RS flipflop using NOR gates*

Output – BEFORE inputs are applied

Output – AFTER inputs are applied

R	S	Q	Q^+	Comment
0	0	0	0	No change (HOLD)
0	1	0	1	SET
1	0	0	0	
1	1	0	X	Indeterminate
0	0	1	1	No change (HOLD)
0	1	1	1	
1	0	1	0	RESET
1	1	1	X	Indeterminate

Figure 6.7 *The RS flipflop: truth table*

The RS (reset–set) bistable or flipflop

The circuit in Figure 6.6 shows the RS flipflop made up from two cross-connected NOR gates. The output from each NOR gate provides one of the inputs to the other NOR gate.

The truth, or state, table in Figure 6.7 shows the final state of the output (Q^+) for both sets of initial output conditions, that is, for $Q = 0$ and $Q = 1$.

Practical Exercise 6.2

To investigate the action of the RS flipflop
For this exercise you will need the following components and equipment:

1 – 74LS02 ic (quad 2-input NOR)
1 – LED (5 mm) and resistor (270 Ω)
1 – +5 V DC power supply

Continued on p. 102

Practical Exercise 6.2 (*Continued*)

Procedure

1 Make up the circuit of Figure 6.8. The pin connection diagram for the 7402 ic is given in Figure 2.13(d).

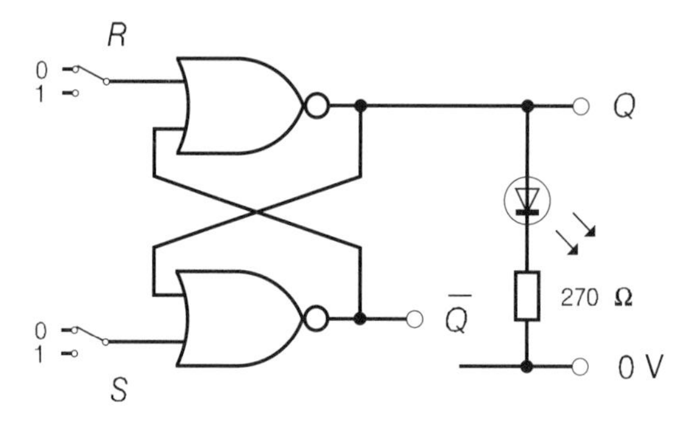

Figure 6.8 *The RS flipflop: circuit for Practical Exercise 6.2*

2 Work through the truth table of Figure 6.9 line by line, adding the appropriate comments. Hence confirm the comments column of Figure 6.7.

3 You should experience some difficulty in getting a consistent Q^+ output for the situation where $R = S = 1$ (the so-called indeterminate state).

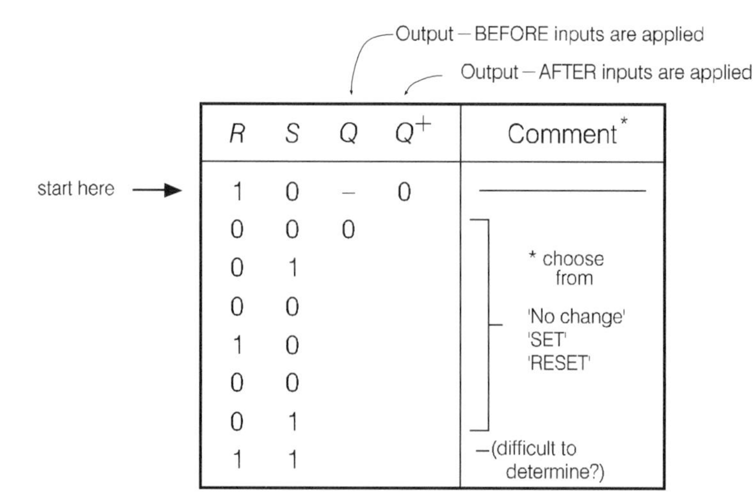

Output – BEFORE inputs are applied (Q)
Output – AFTER inputs are applied (Q^+)

start here →

R	S	Q	Q^+	Comment*
1	0	–	0	———
0	0	0		
0	1			
0	0			
1	0			
0	0			
0	1			
1	1			–(difficult to determine?)

* choose from 'No change' 'SET' 'RESET'

Figure 6.9 *The RS flipflop: truth table for Practical Exercise 6.2*

Important points

- Determining the state of the final output must take into account the previous output as well as the present inputs.
- Q represents the previous state of the output, that is, the state before the present inputs are applied.
- Q^+ represents the state of the output Q after it has responded to the applied input conditions.
- Q and $\overline{Q}$ are always complementary. They can never be at the same logic level at the same time. (Remember the see-saw!)
- For the NOR gate flipflop, the output Q is always taken from the gate with the R (RESET) input.
- The condition $R = 1$, $S = 1$ is called indeterminate because it is impossible to determine what the outputs actually will be. (If you were not convinced from trying it yourself, then read the explanation relating to Figure 6.10.) From the practical point of view, this condition **must not** be allowed to occur.
- Because of the dependency of the output on the present inputs and the previous output, the flipflop (bistable) is a form of memory circuit.

The first two lines of the truth table of Figure 6.7 are now explained in detail, using the diagrams in Figure 6.10.

Line 1

With $R = 0$ and $S = 0$, previous output (Q) from NOR gate A is $Q = 0$.

The feedback from the output (Q) of NOR gate A means that both inputs to NOR gate B are 0. Thus the output ($\overline{Q}$) of NOR gate B is 1.

The feedback from the output ($\overline{Q}$) of NOR gate B means that the inputs to NOR gate A are 0 and 1, giving an output (now called Q^+) from NOR gate A, of 0. There is thus no change in the output Q to Q^+ following the applied input conditions $R = 0$, $S = 0$.

Notice that the final conditions give complementary (opposite) outputs at Q and $\overline{Q}$.

Line 2

With $R = 0$ and $S = 1$, again, $Q = 0$ but now, the feedback from the output (Q) of gate A to the input of gate B means that the output ($\overline{Q}$) of gate B (and thus the input of gate A) will, for the moment, be 0. This appears to contradict the requirement that the outputs must always be complementary. (Read on!)

With both inputs to gate A at 0, the output (Q) will be 1. In other words, there is a change in the state of Q to Q^+ from 0 to 1. This is the SET state, the two outputs **are complementary**, so all is well.

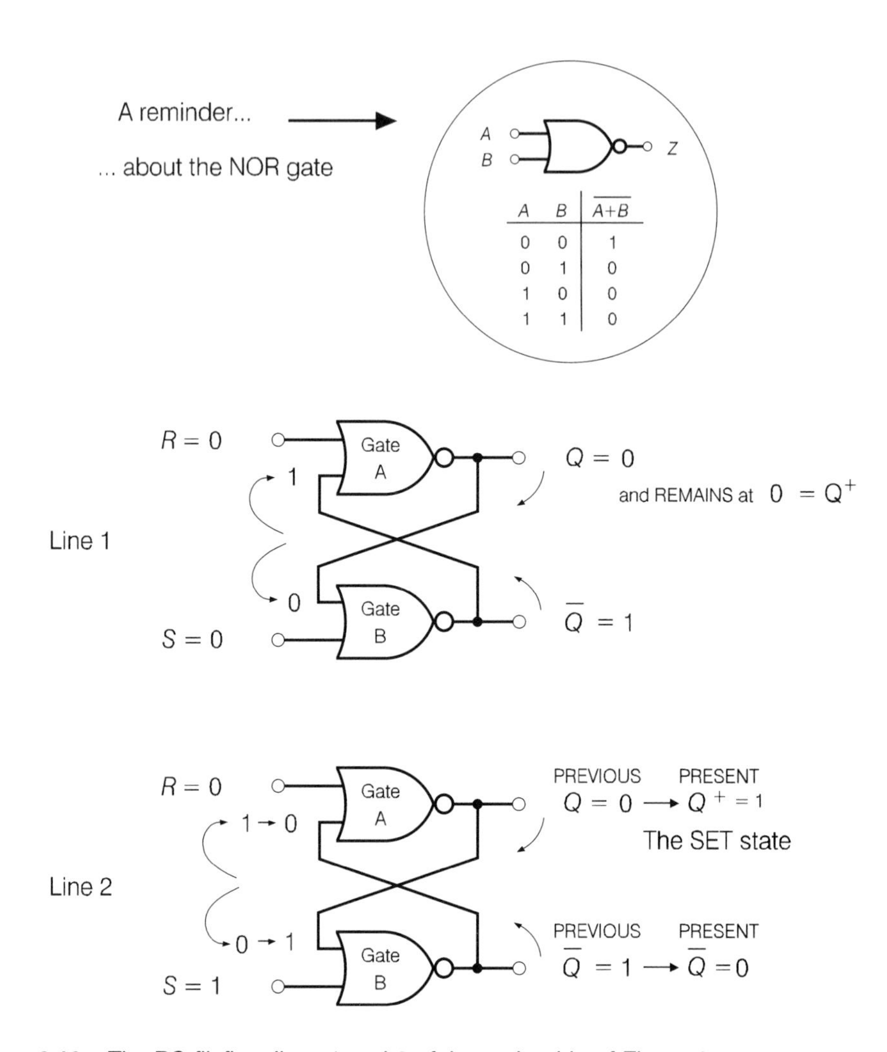

Figure 6.10 *The RS flipflop: lines 1 and 2 of the truth table of Figure 6.7*

The reader is invited to continue the process to complete the exercise.

With $R = S = 1$, a so-called **race condition** develops, with both Q and $\overline{Q}$ attempting to rise or fall simultaneously. Because of the cross-coupling arrangement, they counteract the change in each other. There is a race which ends by one output becoming a 1 ahead of the other – the winner being determined by the characteristics of the individual gates: not a satisfactory state of affairs!

Since the inputs to digital circuits may be rectangular waveforms, it is necessary to be able to understand the so-called **timing diagrams.** Figure 6.11 shows an example, with the explanation provided in tabular form in Figure 6.12.

The RS flipflop can also be made up from two cross-connected NAND gates, as shown in Figure 6.13. Note that for the SET state ($Q = 0$ to $Q^+ = 1$ with $R = 0$, $S = 1$), the output Q is taken from the gate with the S (SET) input, which is the opposite situation to the NOR flipflop. This will then give the correct output conditions for this SET state.

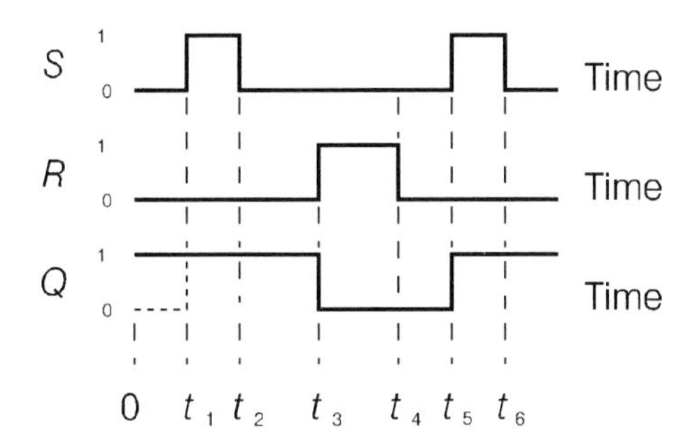

Figure 6.11 *The RS flipflop: timing diagram*

Instant in time	R	S	Q	comments	refer to truth table (Fig.6.8)
0	0	0	1	Q initially at logic 1	
t_1	0	0 to 1	1	Q already at logic 1	
alternatively					
0	0	0	0	Q initially at logic 0	
t_1	0	0 to 1	0 to 1	SET state	line 2
t_2	0	1 to 0	1	No change, HOLD state	line 5
t_3	0 to 1	0	1 to 0	RESET state	line 7
t_4	1 to 0	0	0	No change, HOLD state	line 1
t_5	0	0 to 1	0 to 1	SET state	line 2
t_6	0	1 to 0	1	No change, HOLD state	

Figure 6.12 *The RS flipflop: explanation of the timing diagram*

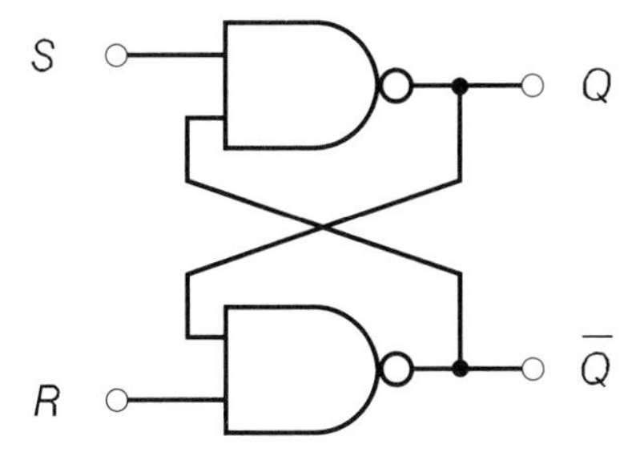

Figure 6.13 *The RS flipflop: using NAND gates*

Before moving on to an improved version, it is worth mentioning (actually as a reminder!) that the RS flipflop described above finds application as a switch debouncer. This was referred to in Chapter 1.

A problem is that the change in state at the output of the above RS flipflop is assumed to occur at the same time as the change in input. However, there could be difficulties due to:

(a) both inputs not changing at the same instant in time
(b) the propagation delay through the system.

The result of this is that the output state would either be unreliable or incorrect.

A solution is to ensure that the operations SET and RESET only occur at certain specified times. This is achieved by the use of trigger or clock pulses.

The clocked RS flipflop

The circuit diagram is shown in Figure 6.14.

The purpose of the clock input is to provide a signal in pulse form to enable the data at the input to be transferred to the output. A clocked flipflop will change state only when the clock pulse is applied. The actual change of state or clocking can be made to occur when the clock pulse is going either from 0 to 1, called **positive edge** triggering, or from 1 to 0 (**negative edge** triggering). This is shown in Figure 6.15. The choice of which to use is decided by the flipflop type, the symbols for which are shown in Figure 6.16.

Important points

- Change of output state occurs only when clock pulse is applied.
- Before the clock pulse is applied, the inputs can be set without causing any change in output state.
- The advantage of clocking is that all changes occur at the same time or, as it is called, in synchronism. This is particularly important when several flipflops are connected together. A clocked flipflop is thus a synchronous flipflop.
- The flipflop operates in step with the clock – a very important characteristic in, for example, calculators and computers, where the actions must occur in a certain order.

Action of the clocked flipflop

Refer again to the circuit in Figure 6.14, together with a partial truth table in Figure 6.17. The complete truth table for the clocked flipflop is given in Figure 6.18.

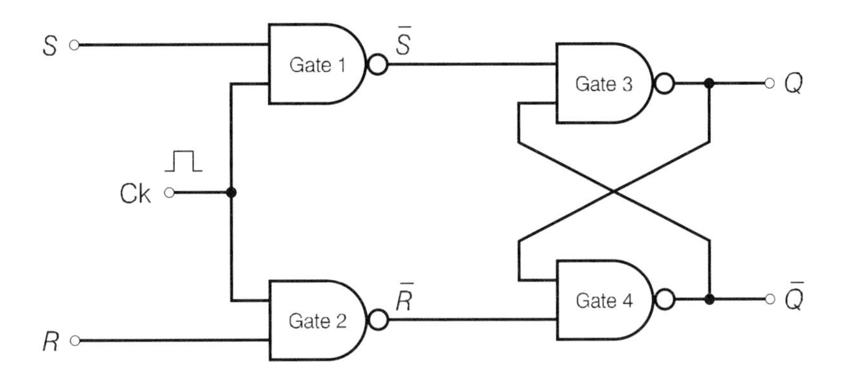

Figure 6.14 *The clocked RS flipflop using NAND gates*

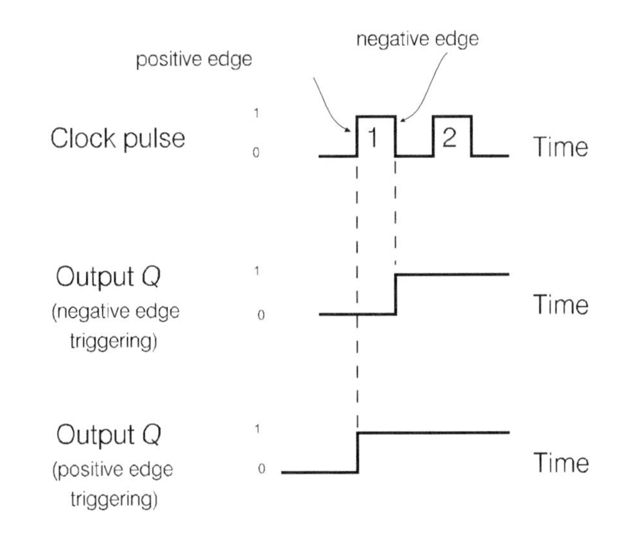

Figure 6.15 *Negative and positive edge triggering*

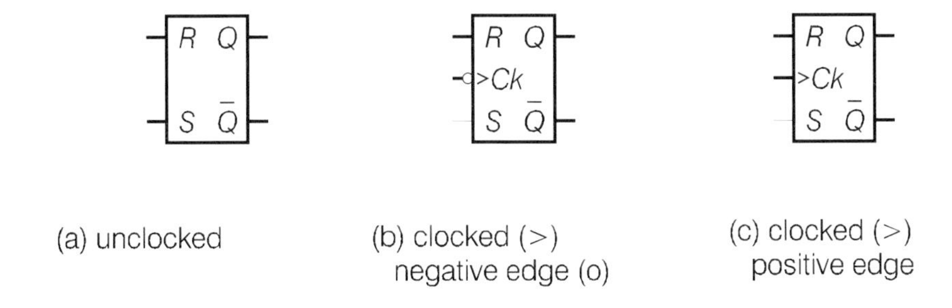

Figure 6.16 *The RS flipflop: circuit symbols clocked and unclocked*

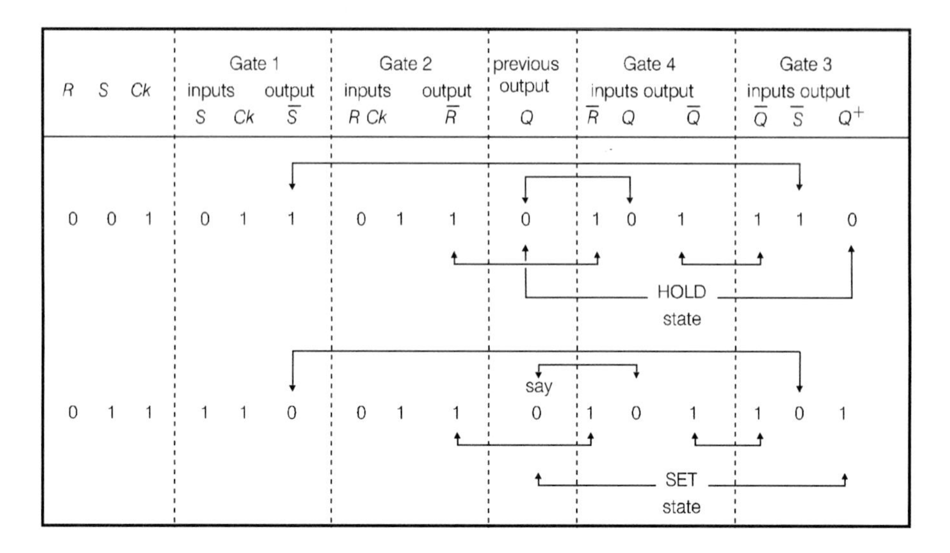

R	S	Ck	Gate 1 inputs S	Gate 1 inputs Ck	Gate 1 output $\overline{S}$	Gate 2 inputs R	Gate 2 inputs Ck	Gate 2 output $\overline{R}$	previous output Q	Gate 4 inputs $\overline{R}$	Gate 4 inputs Q	Gate 4 output $\overline{Q}$	Gate 3 inputs $\overline{Q}$	Gate 3 inputs $\overline{S}$	Gate 3 output Q^+
0	0	1	0	1	1	0	1	1	0	1	0	1	1	1	0
									HOLD state						
0	1	1	1	1	0	0	1	1	say 0	1	0	1	1	0	1
									SET state						

Figure 6.17 *The clocked RS flipflop: partial truth table*

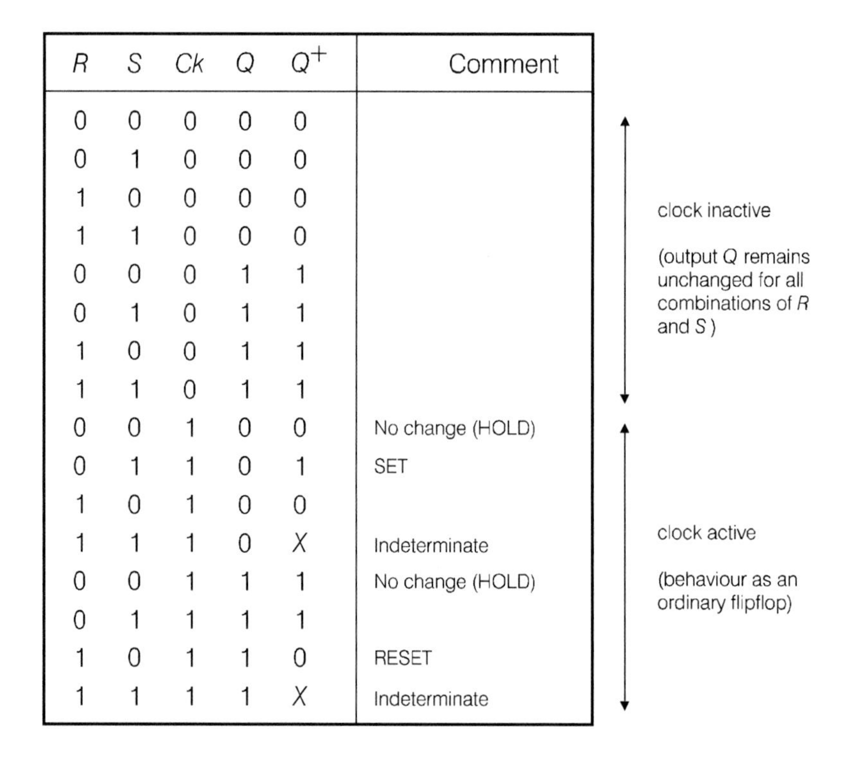

R	S	Ck	Q	Q^+	Comment	
0	0	0	0	0		clock inactive (output Q remains unchanged for all combinations of R and S)
0	1	0	0	0		
1	0	0	0	0		
1	1	0	0	0		
0	0	0	1	1		
0	1	0	1	1		
1	0	0	1	1		
1	1	0	1	1		
0	0	1	0	0	No change (HOLD)	clock active (behaviour as an ordinary flipflop)
0	1	1	0	1	SET	
1	0	1	0	0		
1	1	1	0	X	Indeterminate	
0	0	1	1	1	No change (HOLD)	
0	1	1	1	1		
1	0	1	1	0	RESET	
1	1	1	1	X	Indeterminate	

Figure 6.18 *The clocked RS flipflop: complete truth table*

Gates 1 and 2 are the input gates and they inhibit the RS inputs, that is, prevent them from taking effect, until the clock input (*Ck*) becomes active (logic 1). They also ensure that all changes take place in synchronism with the system clock.

Important points

- With the clock inactive ($Ck = 0$) from lines 1 to 8, there is no change in the output state (Q to Q^+) for all input combinations of R and S. In other words, nothing happens!
- Changes in the output state occur only when the clock is active ($Ck = 1$) from lines 9 to 16, when the behaviour is that of the RS flipflop met earlier.

A waveform (timing) diagram example is given in Figure 6.19. The essential feature of clocked flipflops is that changes in the output Q will take place only when the clock pulse is active – for this example, when the clock pulse is going from logic 0 to logic 1 – and with the correct set of circumstances for R and S. Thus changes in Q take place only at times t_1, t_2 and t_3.

It would be helpful if the timing diagram was read in conjunction with the truth table of Figure 6.18.

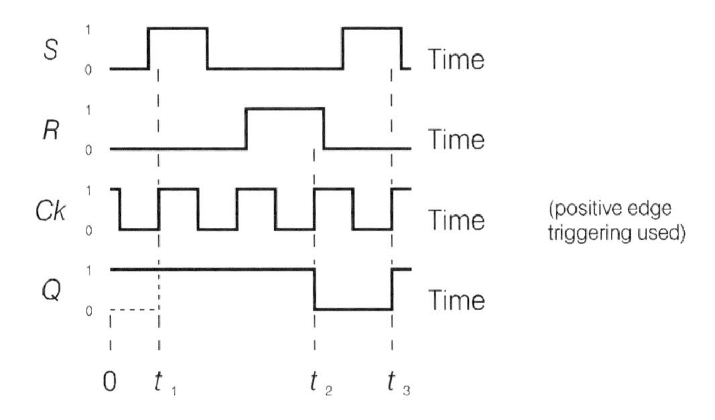

Figure 6.19 *The clocked RS flipflop: timing diagram*

Questions

6.1 (a–d) For each set of waveforms given in Figure 6.20 draw the resulting output waveform *Q*.

Continued on p. 110

Questions (*Continued*)

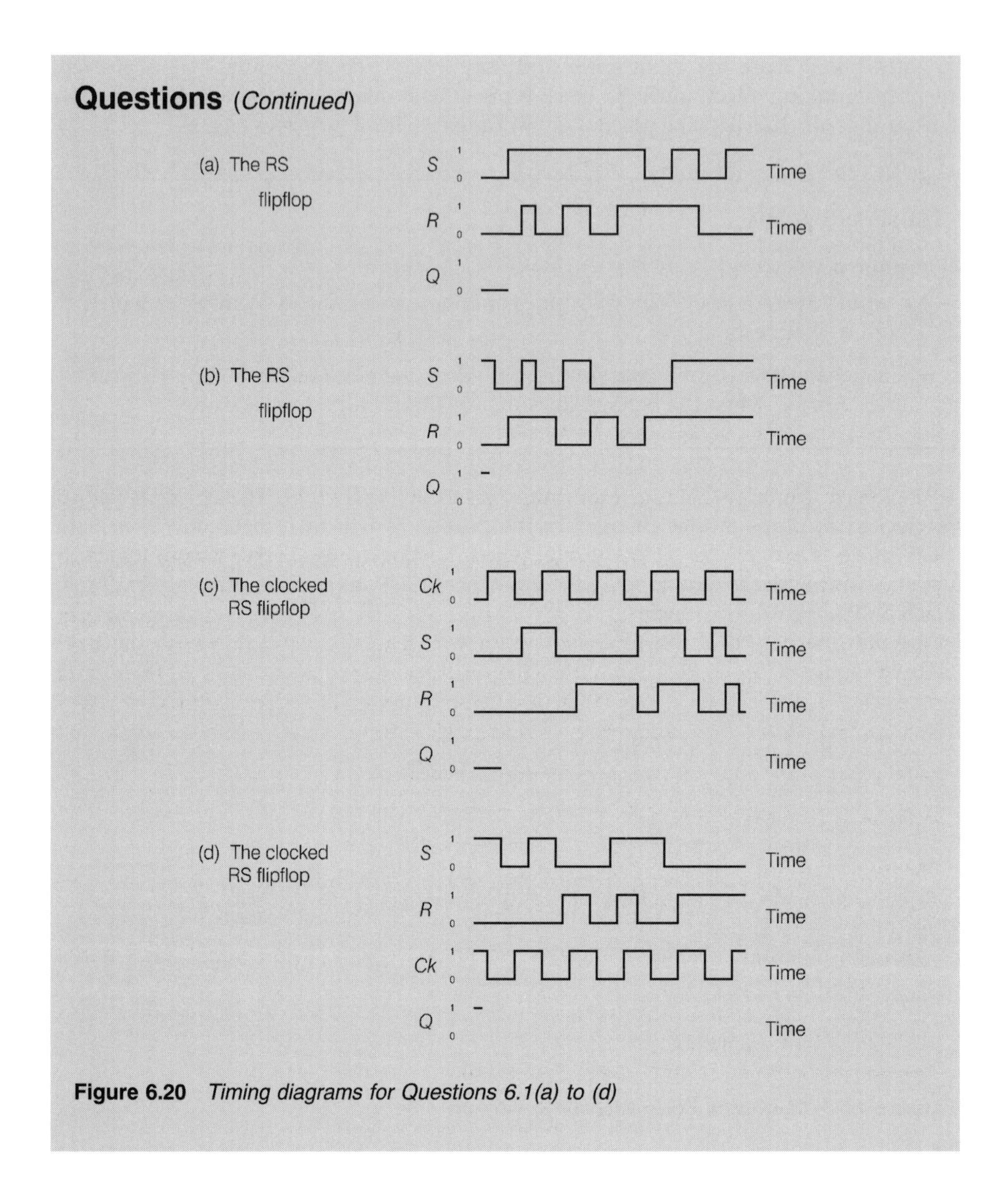

Figure 6.20 *Timing diagrams for Questions 6.1(a) to (d)*

The D-type flipflop

This is another example of a clocked flipflop. The D-type is used for temporary data storage, the letter D standing for data or delay, as will be evident later. The circuit symbol diagram is shown in Figure 6.21. Note that the circuit symbol shows both a preset (SET) and clear (RESET) pin.

Preset. As the name suggests, this will take the output to a preset or predetermined level. A logic 0 at the preset input will take the Q output to logic 1. This is known as

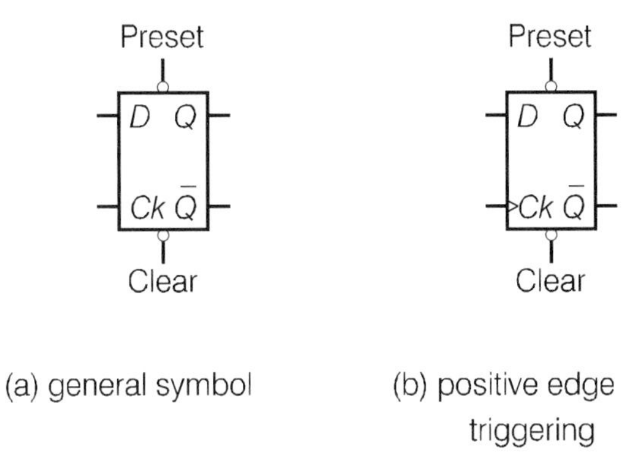

Figure 6.21 *The D flipflop: circuit symbols*

active low since the required action of the preset input occurs when its logic level is 0 (low).

Clear. This will take the Q output to logic 0 when a logic 0 is applied to the clear input. This again is known as **active low**.

Since both the preset and clear inputs will override the data inputs – the flipflop can be preset or cleared at any time – they are known as **asynchronous inputs**.

The basic action of the D-type is that the output (Q) will take the same logic state as the data input (D) whenever the clock pulse is applied (which, for positive edge triggering, is when the clock pulse goes from logic 0 to logic 1).

Practical Exercise 6.3

To investigate the action of the D flipflop
For this exercise you will need the following components and equipment:

1 – 74LS74 ic (dual D-type, using positive edge triggering, with preset and clear inputs)
1 – LED (5 mm) and resistor (270 Ω)
1 – +5 V DC power supply
1 – pulse generator (1 kHz – see Figure 1.5)
1 – double beam cathode ray oscilloscope

Procedure (a)

For the whole of this exercise the preset input will not be used, and it should therefore be permanently connected to logic 1.

1 Connect up the circuit shown in Figure 6.22. The pin connection diagram for the 7474 ic is given in Figure 6.23.

Continued on p. 112

Practical Exercise 6.3 *(Continued)*

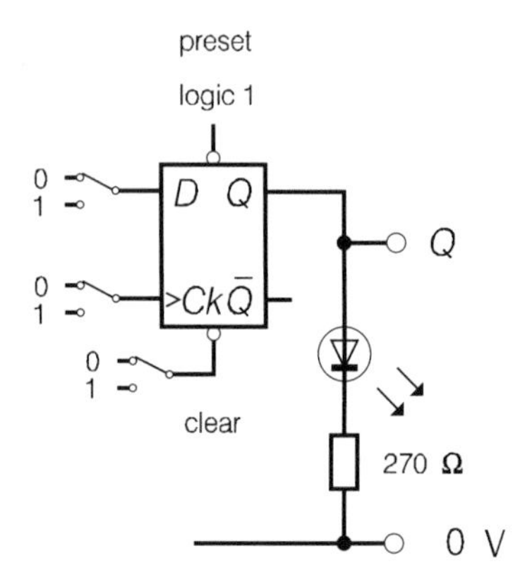

Figure 6.22 *The D flipflop: circuit for Practical Exercise 6.3(a)*

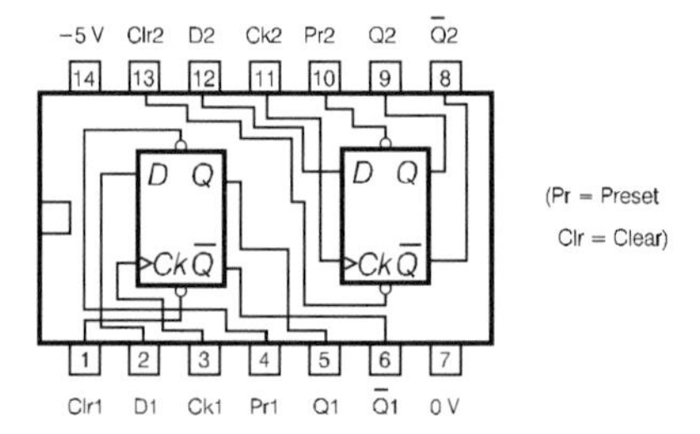

Figure 6.23 *The 7474 dual D flipflop: pin connections*

2 Make sure that the output *Q* is initially at logic 0 by connecting the clear input momentarily to logic 0, after which it should be connected to logic 1.
3 With the clock input at logic 0, connect a logic 1 to the *D* input and notice that there is no change in the state of the *Q* output.
4 Leave the *D* input at logic 1, connect a logic 1 to the clock input and notice that the *Q* output goes to logic 1 (that is, to the same logic state as the *D* input).
5 Return the clock input to logic 0 and notice that the *Q* output remains at logic 1.
6 Connect a logic 0 to the *D* input and notice that when the clock input goes from logic 0 to logic 1, the *Q* output goes to logic 0 (the same logic state as the *D* input).
7 Check this action further by working through the sequence shown in the table of Figure 6.24 in the order given.

Procedure (b)

1 Modify the circuit to that of Figure 6.25(a).

INPUTS				OUTPUT	
Preset	Clear	*D*	Clock	*Q*	Comment
0	1	*X*	*X*	1	Asynchronous SET
1	0	*X*	*X*	0	Asynchronous RESET
0	0	*X*	*X*	1	Prohibited
1	1	1	0→1	1	SET
1	1	0	0→1	0	RESET

(X = 'doesn't matter')

Figure 6.24 *The D flipflop: truth table*

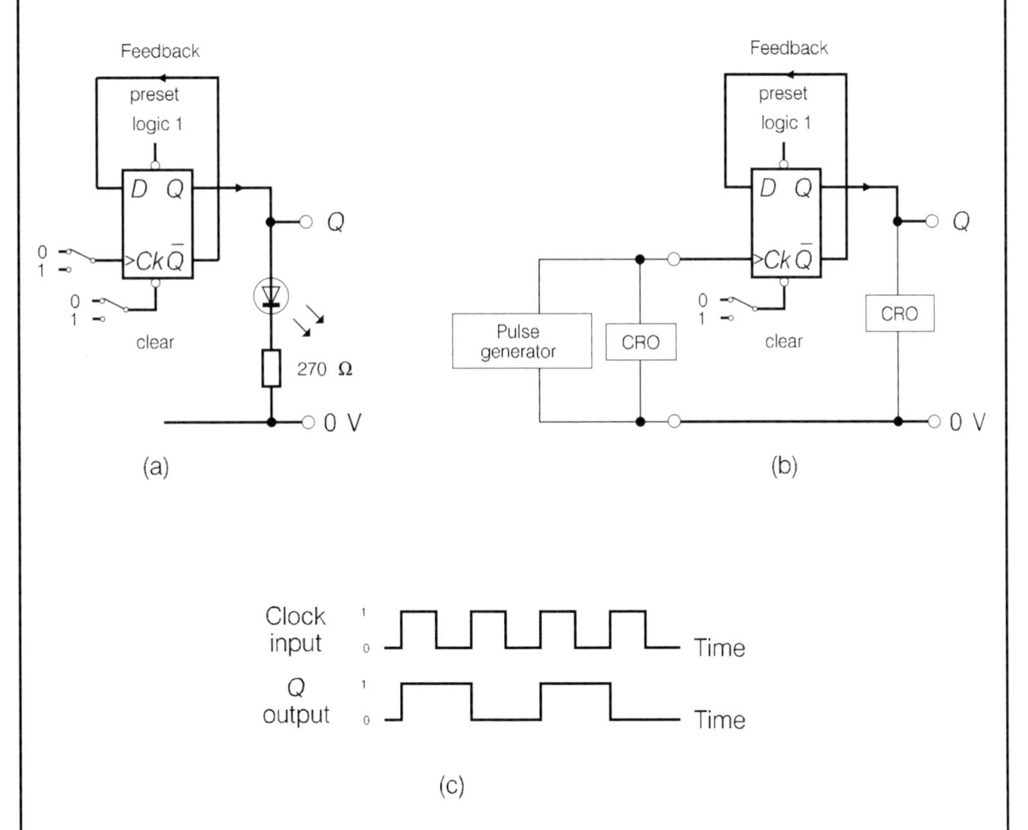

Figure 6.25 *The D flipflop with feedback: circuits and waveforms for Practical Exercise 6.3(b)*

2 Apply two complete clock pulses (two on/offs) and notice that the *Q* output consists of one on/off, that is, one complete pulse. The D-type flipflop is said to

Continued on p. 114

Practical Exercise 6.3 (*Continued*)

be **dividing by two**. This action forms the basis of the binary counter, which is dealt with in the next chapter.

3 If the pulse generator and cathode ray oscilloscope are available, modify the circuit to that of Figure 6.25(b). The waveforms should be as shown in Figure 6.25(c). The clear (RESET) action can be demonstrated by setting the clear input to logic 0 (at any time), whereupon the Q output should then become logic 0.

Important points

Since it is customary to use devices other than the D-type for binary counting, a brief summary of the action in procedure (b) above is appropriate.

- The use of **feedback** from the $\overline{Q}$ (inverted) output to the D input: on each positive-going clock pulse, the inverse (opposite) of the Q output state, ie. $\overline{Q}$, is fed back to the D input. This causes the Q output to change to this inverse state on the next clock pulse. Thus for a series of input clock pulses of a particular frequency, there will be a series of Q output pulses, at one-half of this frequency.
- The term **delay flipflop** describes what happens to the information (data) at the input D. This data arrives at the output Q following the application of one clock pulse and there is thus a one-bit time delay.
- The advantage of the D flipflop over the RS flipflop is that the D-type has only one external data input, thus avoiding the indeterminate situation ($R = S = 1$) of the RS type. However, for the D flipflop, the situation where both preset and clear are logic 0 is naturally prohibited!

The latch

In the brief description of the latch at the beginning of this chapter, we saw that following a given input, the output condition remained in its **set** state until a **reset** input signal was applied.

A latch is therefore a temporary digital storage (memory) device, with a typical example being the 7475 4-bit bistable ic, consisting of four D-type flipflops and two enable (or clock) inputs, normally connected together.

With the **clock** (enable) input at **logic 1**, data present at the D input is transferred to the Q output, and **any changes** in the D state will be **followed** at the Q output. In this state the device is known as a **transparent latch**.

When the **clock** (enable) changes to **logic 0**, the Q output will remain at its state immediately before the change in clock input, and **will not respond** to changes in the D input. It is said to be **data latched**.

One application for such a device is in counting, where it is often necessary to freeze the display in order to check the total count while counting is allowed to continue.

Important points

To recognize the operational differences between a D flipflop (7474 ic) and a prescribed D latch (7475):

For the D flipflop:

- There is one data input (D), a clock input and complementary outputs (Q and $\overline{Q}$). It is therefore a 1-bit storage device.
- Data is transferred from the D input to the Q output **only** on the **edge** of a clock pulse.

For the D latch:

- Having four inputs and outputs, it is a 4-bit storage device.
- The Q outputs will follow the D inputs as long as the clock (enable) input remains at logic 1 (a transparent latch). Otherwise, the Q outputs will remain at their previous level, even if the D inputs change (a data latch).

Single-bit memory with write and read facility

The use of the D flipflop as a memory unit is shown by the circuit diagram in Figure 6.26.

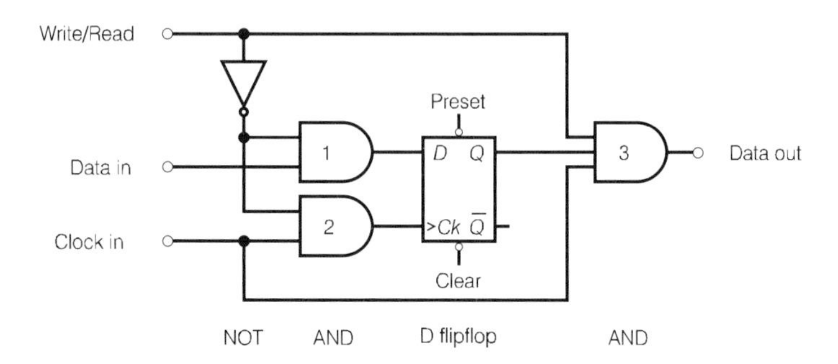

Figure 6.26 *The D flipflop as a single-bit memory cell with write and read facility*

The system has two stages of operation, namely

(i) A 'write' stage in which the input data is taken into memory, in preparation for
(ii) The 'read' stage in which the stored data is delivered to the output.

At the write/read input

- The **Write** position is identified by logic 0
- The **Read** position is identified by logic 1.

Now to consider the two-stage operation in more detail.

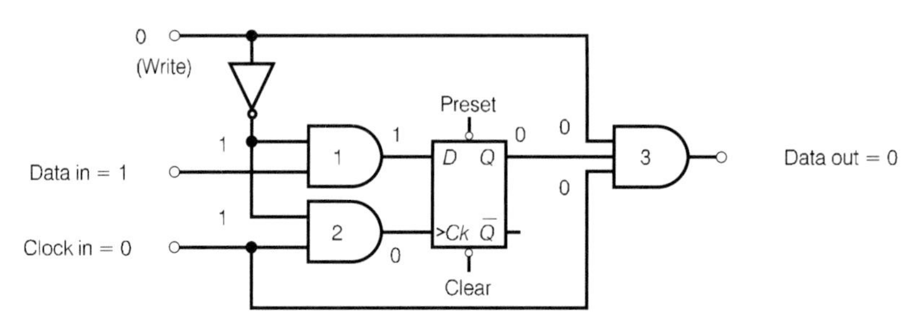

Input data (either 1 or 0) is transferred through AND gate 1 to the D input of the flipflop

(a) Write: Clock = 0

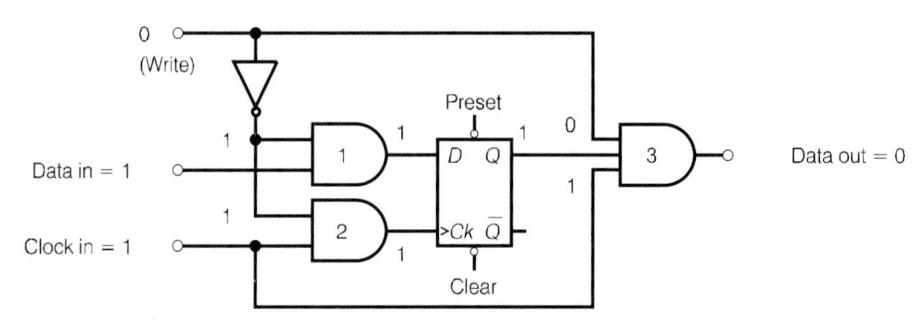

When the clock input = 1, the D input is clocked into the Q output and held there

(b) Write: Clock = 1

Figure 6.27 *The D flipflop as a single-bit memory cell with write and read facility: the 'write' sequence*

(i) Logic 0 = **Write** (Figure 6.27)
- Output of AND gate 1 will **follow** the data input. (Data input = 1, output of AND gate 1 = 1, D input = 1)
- Following one clock pulse, the D input will be transferred to the Q output ($Q = 1$) and held there.
- Output of memory unit (AND gate 3) will be logic 0.

(ii) Logic 1 = **Read** (Figure 6.28)
- Output of AND gate 1 and AND gate 2 will be logic 0. (Q output remains at logic 1 from the previous 'write' operation)
- When the clock input to AND gate 3 is logic 1, the output from this gate will be the same as the Q output (in this example, logic 1).

It would now be helpful to verify this operation in practice, and extend it to a 4-bit memory unit.

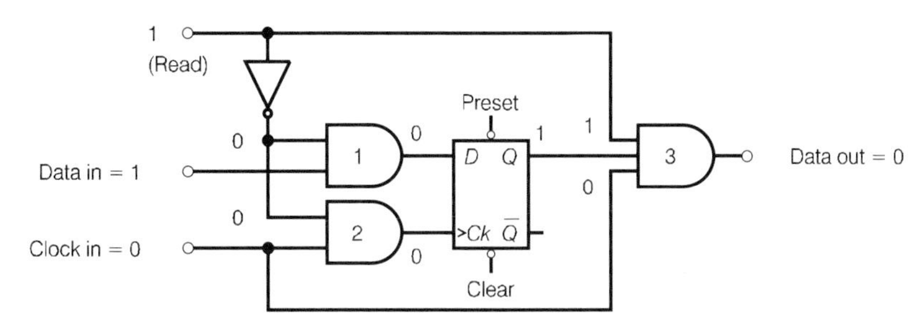

No output from either AND gate 1 or 2. *Q* output held at 1

(a) Read: Clock = 0

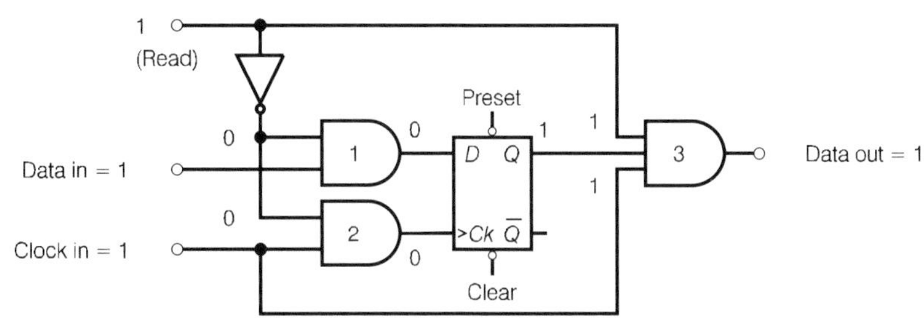

When the clock input = 1, the data output will be 1

Note:
When the clock input returns to 0, so will the data output, but the output at *Q* remains unchanged

(b) Read: Clock = 1

Figure 6.28 *The D flipflop as a single-bit memory cell with write and read facility: the 'read' sequence*

Practical Exercise 6.4

To investigate a single-bit memory unit

For this exercise you will need the following components and equipment:

1 – 74LS04 ic (hex inverter) (NOT gate)
1 – 74LS08 ic (quad 2-input AND gate)
1 – 74LS11 ic (triple 3-input AND gate)
1 – 74LS74 ic (dual D flipflop)
4 – LED (5 mm) + series resistor (270 Ω)
1 – +5 V DC power supply

Continued on p. 118

Practical Exercise 6.4 (*Continued*)

Procedure

1 Connect up the circuit shown in Figure 6.29. The pin connections for the 7404 and 7408 ic's are given in Figure 2.13, the 7411 ic in Figure 6.30 and the 7474 ic in Figure 6.23.
The preset input should be permanently connected to logic 1.

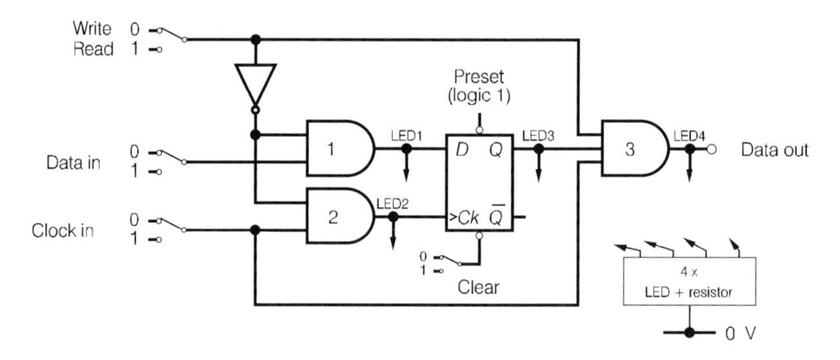

Figure 6.29 *The D flipflop as a single-bit memory cell: circuit for Practical Exercise 6.4*

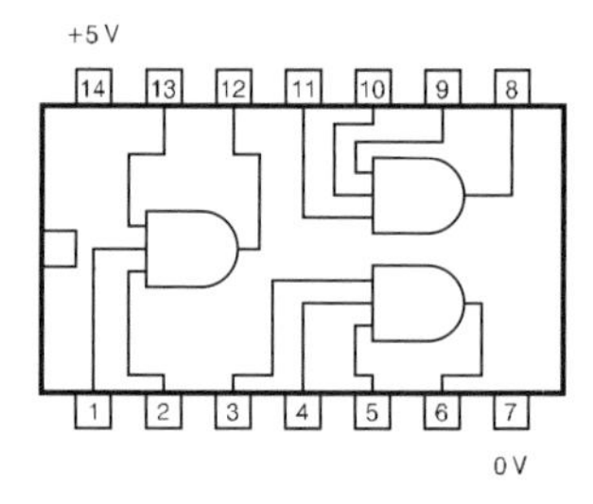

Figure 6.30 *The 7411 triple 3-input AND gate: pin connections*

2 Make sure that the *Q* output is initially at logic 0 by connecting the clear input momentarily to logic 0, after which it should be connected to logic 1.
3 Set the write/read input to 'write'.
4 With the data and clock inputs as in Figure 6.27(a), only LED 1 (*D* input) should be lit.
5 When the clock input is switched from logic 0 to logic 1 (positive edge triggering), LEDs 1, 2 and 3 should be lit. The *Q* output takes the same state as the *D* input, and remains at that state even though the clock input may change.
Note that at this point, the data (information) is being stored at the *Q* output.
6 Set the write/read input to 'read'.
7 With the data and clock inputs as in Figure 6.28(a), only LED 3 (*Q* output) should be lit.

8 When the clock input is switched from logic 0 to logic 1, LEDs 3 and 4 (data output) should be lit.

The data output takes the same state as the Q output, and the Q output remains in that state even though the clock input may change.

Conclusion

1 Using Figure 6.26 as a guide, draw the circuit diagram of a 4-bit memory unit.

A Practical Exercise involving the use of the 7475 latch will be found in Chapter 10.

The JK flipflop

This is a versatile, widely used and universal flipflop, possessing the features of all the other types. The letters J and K have no literal meaning, but are compared to the RS flipflop, J with S (SET) and K with R (RESET).

There is additional circuitry which prevents the JK flipflop being forced into the indeterminate state that occurs when S and R are both logic 1.

It is a **clocked** flipflop which may also have a preset and clear input. The J, K and clock inputs are **synchronous**, whereas the preset and clear are **asynchronous**, that is, they can be applied at any time and will override any existing state.

Figure 6.31 shows the circuit symbols for the JK flipflop.

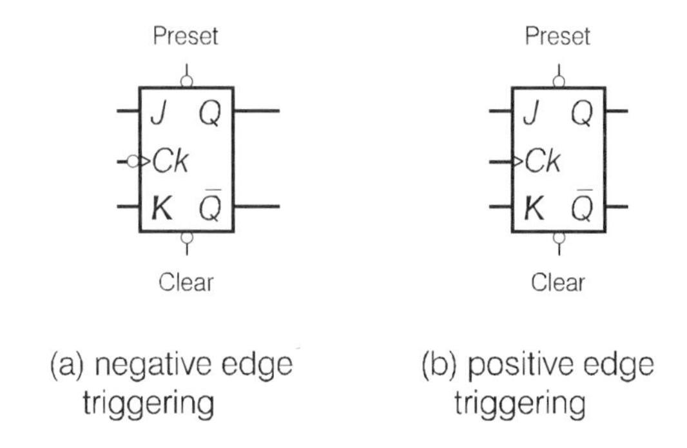

Figure 6.31 *The JK flipflop: circuit symbols*

Practical Exercise 6.5

To investigate the action of the JK flipflop

For this exercise you will need the following components and equipment:

Continued on p. 120

Practical Exercise 6.5 (*Continued*)

1 – 74LS76 ic (dual JK flipflop with preset and clear active low, and negative edge triggering)
1 – 7LED (5 mm) and resistor (270 Ω)
1 – +5 V DC power supply
1 – pulse generator (1 kHz – see Figure 1.5)
1 – double beam cathode ray oscilloscope

Procedure (a)

1 Connect up the circuit in Figure 6.32. The pin connection diagram for the 7476 is shown in Figure 6.33. Notice that both the preset and clear inputs are connected to logic 1 since they are not needed in this exercise.

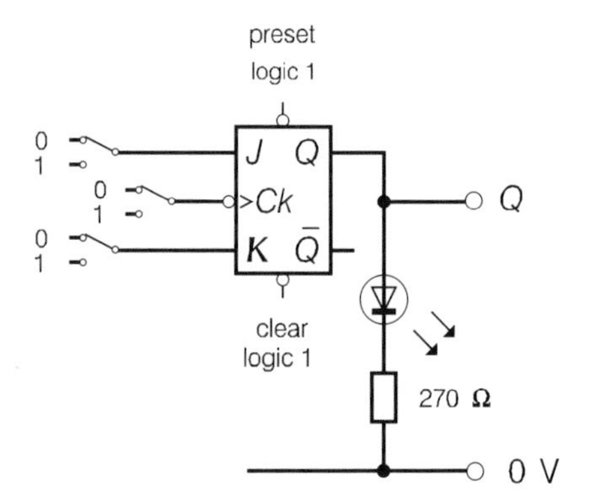

Figure 6.32 *The JK flipflop: circuit for Practical Exercise 6.5*

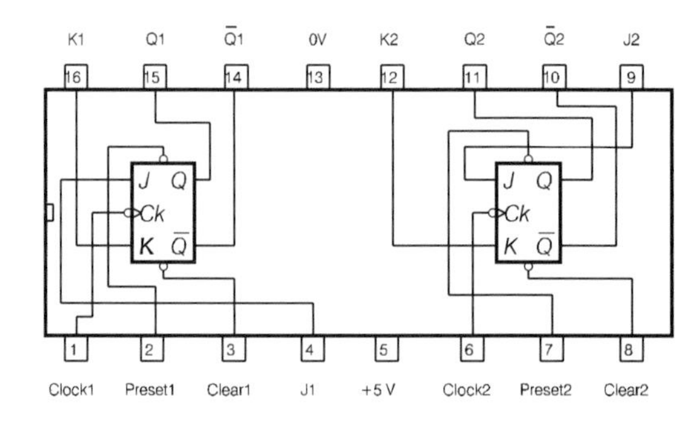

Figure 6.33 *The 7476 dual JK flipflop: pin connections*

2 Work through the input sequence shown in the truth table of Figure 6.34, in the order given, and hence complete the output columns.

INPUTS		OUTPUT		Comments
		Before Clock pulse	After Clock pulse	(Hold, Set, Reset, etc)
J	*K*	(*Q*)	(Q^+)	
0	1	—	0	——
0	0	0		Hold (No change)
1	0			SET
0	0			Hold
0	1			RESET
1	1			Toggle
1	1			Toggle
1	1			Toggle
1	1			Toggle

Figure 6.34 *The JK flipflop: truth table for Practical Exercise 6.5*

3 Check that your entries in the truth table agree with the summary of JK flipflop action given below.

If a pulse generator and oscilloscope are available you can now continue with the next part of the exercise.

Procedure (b)

1 Connect the *J* and *K* inputs permanently to logic 1 (the present and clear remain at logic 1).
2 Connect the generator to the clock input (*Ck*), and the oscilloscope inputs to the clock input and *Q* output respectively.
3 The resulting waveforms should resemble those shown for *Ck* and *Q* in Figure 6.35. Notice that changes to the *Q* output state occur only when the clock pulse goes from 1 to 0 (the **negative edge**).

Summary of JK flipflop action

Inputs	*Action at output Q*	*Comments*
$J=0, K=0$	No change	HOLD state
$J=1, K=0$	Logic 1	SET state
$J=0, K=1$	Logic 0	RESET state
$J=1, K=1$	Reverses after each clock pulse	TOGGLE state

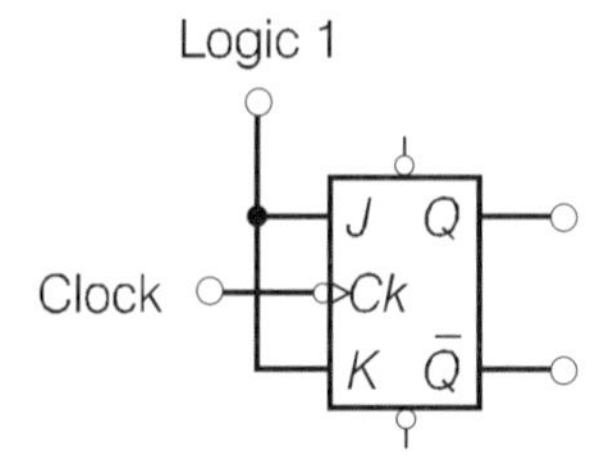

(a) The toggling connections

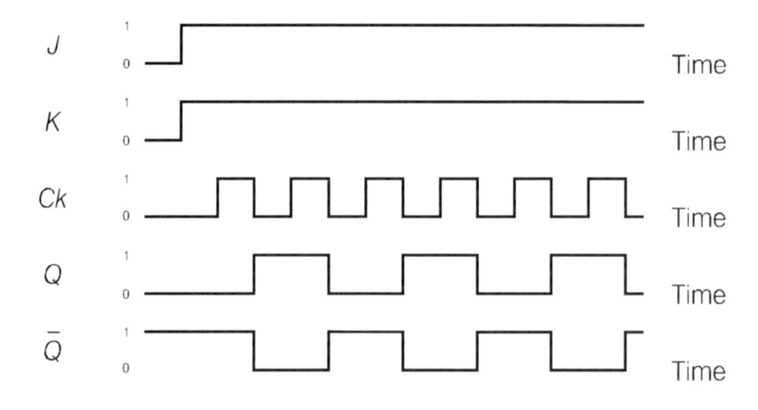

(b) The timing diagram, assuming negative edge triggering

Figure 6.35 *The JK flipflop in a toggling mode*

Important points

- The condition $J = 1$, $K = 1$ (a forbidden state with the RS flipflop) is allowed here. Notice that the output changes every time that both inputs J and K are logic 1. This **toggling action** plays a very important part in the application of JK flipflops.
- From the last four lines of the table in Figure 6.34 (when $J = K = 1$), notice that the output state changes twice for the four input clock pulses. This toggling mode (dividing by two) is illustrated by the waveforms in Figure 6.35.
  Toggling action forms the basis of the binary counter, dealt with in Chapter 7.
- The use of a clock pulse does away with the uncertainties of the order in which the inputs are applied.

Timing diagrams again form an important part of the understanding of this type of flipflop. An example is given in Figure 6.36.

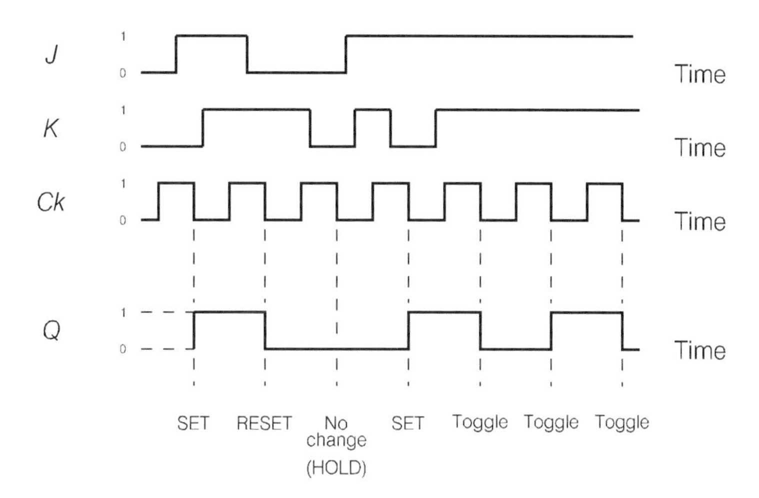

Figure 6.36 *The JK flipflop: timing diagram example*

Problems associated with the JK flipflop

In order to achieve the above conditions, in particular the elimination of the indeterminate state, the construction of the JK flipflop entails having feedback from output to input. Thus, if the Q output changes state before the end of the clock pulse, the input conditions will change. The effect will be that the output Q will oscillate between logic 1 and logic 0 for the duration of the clock pulse, and at the end of the clock pulse the output state is unpredictable.

To avoid this state of affairs, the time for the clock pulse should be short compared with the switching time (propagation delay) of the flipflop. With even faster-switching modern ic's this condition is difficult to meet. The above problem is known as the **race-round condition** and has led to the development of the master–slave flipflop.

The master–slave JK flipflop

Diagrams to illustrate the basic principle of the master–slave flipflop are shown in Figure 6.37.

The clock pulse controls when the switches S_1 and S_2 are open. The arrangement (Figures 6.37(a) and 6.37(b) respectively) is that when S_1 is open, S_2 is closed, and when S_2 is open, S1 is closed. Figure 6.37(c) shows that S_1 will open on the **leading edge** of the clock pulse, causing the master flipflop to be **enabled** (to operate) and S_2 will open on the **trailing edge,** thus enabling the slave.

Important points

- With S_1 open (and S_2 closed) the (input) data is clocked into the input of the enabled master while the slave is disabled.

Continued on p. 124

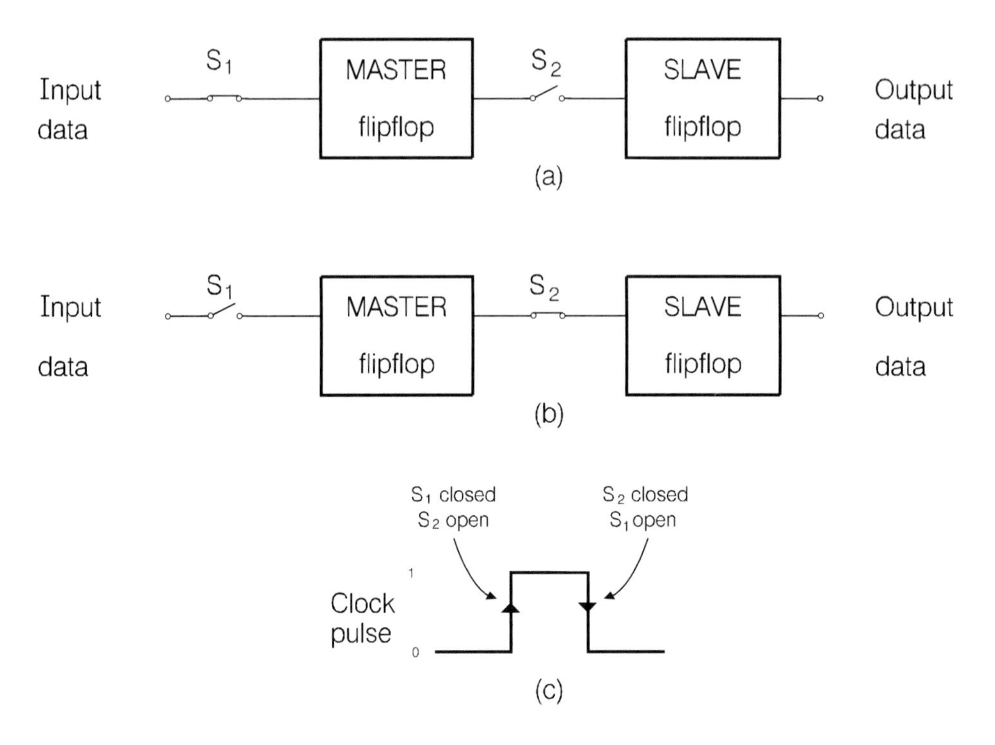

Figure 6.37 *The master–slave flipflop: basic principle*

Important points (*Continued*)

- With S_2 open (and S_1 closed) the master is disabled while the output of the enabled slave follows the output of the master.
- The JK master–slave flipflop uses the entire pulse (that is, both positive and negative edges) in order to transfer data. The master output changes on the positive edge and the slave output on the negative edge of the input clock pulse. The slave output, of course, is the actual output of the device, which is thus **negative edge triggered.**

The waveforms are shown in Figure 6.38.

If you wish to make up a master–slave flipflop using individual ic's, try the next practical exercise. Before you start, a word of caution: there are master–slave flipflops available in ic form, the 7476 being an example. You will see that Practical Exercise 6.6 uses this ic, which operates with negative edge triggering, thus agreeing with the final statements in the points above. As a result of this, you will find that while the principle of operation described is correct (in that it takes one pulse-width to transfer data), there will be a difference – namely that the master output will change on the negative edge and the slave output on the positive edge of the input clock pulse.

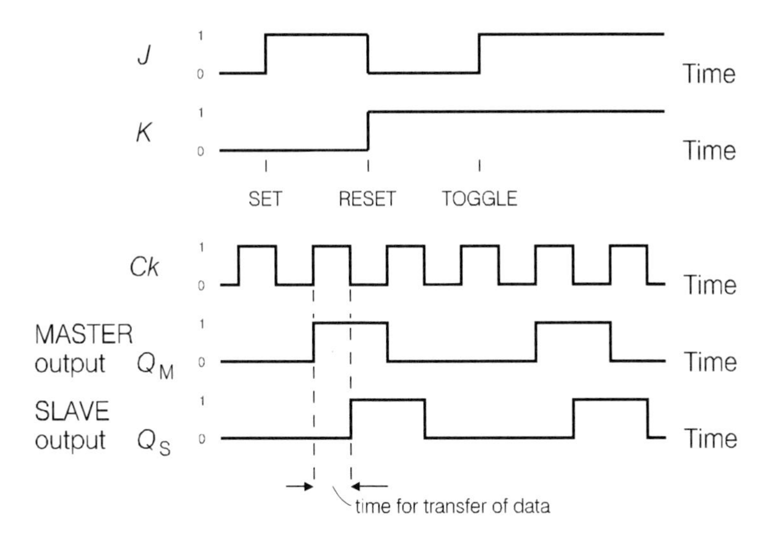

Figure 6.38 *The master–slave flipflop: timing diagrams*

Practical Exercise 6.6

To investigate the action of the Master–Slave flipflop
For this exercise you will need the following components and equipment:

1 – 74LS76 (dual JK flipflop)
1 – 74LS08 (quad 2-input AND)
1 – 74LS11 (triple 3-input AND)
1 – 74LS04 (hex inverter)
2 – LED (5 mm) and resistor (270 Ω)
1 – pulse generator (1 kHz – see Figure 1.5)
1 – double beam cathode ray oscilloscope

The circuit diagram of a master–slave flipflop is shown in Figure 6.39.

Procedure

1 Connect up the circuit in Figure 6.39. The pin connection diagrams for the various ic's will be found as follows:

7476 – Figure 6.33
7408 – Figure 2.13(a)
7411 – Figure 6.30
7404 – Figure 2.13(e)

Continued on p. 126

Practical Exercise 6.6 (*Continued*)

2 Connect the external *J* and *K* inputs to logic 1. (The flipflop will thus be in its toggling mode.)

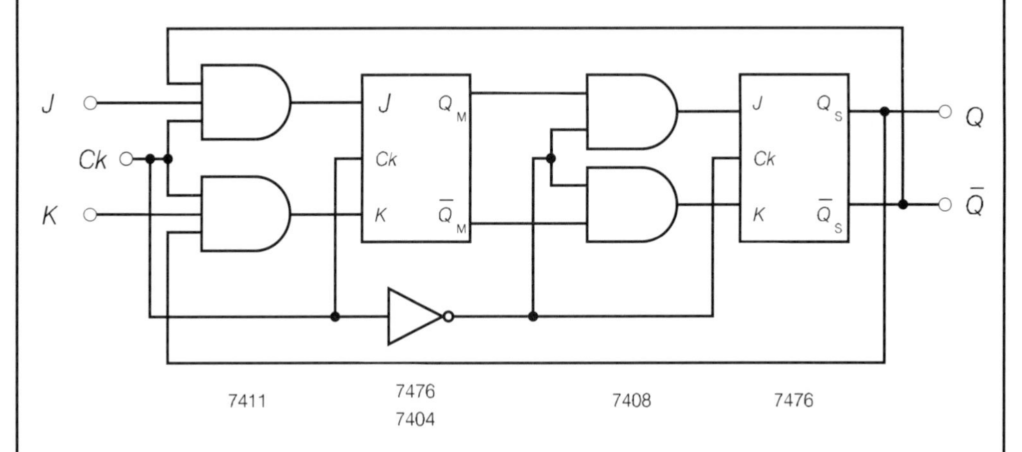

Figure 6.39 *The master–slave flipflop: circuit for Practical Exercise 6.6*

3 For a succession of clock pulses, confirm that output Q_M changes to logic 1 on the negative edge of the clock pulse and output Q_S to logic 1 on the positive edge.

If the 1 kHz square-wave generator and the oscilloscope are available, their use will provide a better visual indication of the operation of the flipflop. Draw the time-related waveforms, one underneath the other, showing the clock pulse, the master output and the slave output, remembering the comments made immediately before this exercise.

Important point

- The output of the M–S flipflop is fully predictable for all combinations of *J* and *K* and for any width of clock pulse.

Questions

6.2 Explain, with reference to the truth table for the JK flipflop, the situation known as toggling. What limitation does the JK flipflop have?

6.3 Describe briefly the operation of the master–slave flipflop and explain its advantage over the JK flipflop.

6.4 Explain the meaning of the term 'active low' as applied to the present and clear inputs of flipflops.

6.5 A certain flipflop uses clear and present inputs that are active low. What will be the state of the Q output if $J = 1$, $K = 0$, for the following conditions:
(a) clear = 0, preset = 1
(b) clear = 1, preset = 1
(c) clear = 1, preset = 0?

6.6 If in this same flipflop the clear and present are not required, to which logic level should they be connected?

6.7 Explain the difference in operation between a D flipflop and a D latch.

6.8 The TTL ic 7474 is known as a dual D-type, positive-edge triggered flipflop, with preset and clear. Using similar terminology, provide the appropriate description for the following ic's:

74107, 74174, 74279.

6.9 Make up a table in order to show the differences in the important characteristics for the following D flipflops:

7474, 74LS74, 74ALS74, 74AC74, 74HC74 and 4013B.

Consider the characteristics **maximum clock frequency, average propagation delay** and **average supply current.**

In addition, note the effect of a change in supply voltage on the propagation delay for the CMOS type 4013B.

7 Sequential systems 2: Counters and shift registers

We have seen that the bistable or flipflop is a divide-by-two device, which means that there will be one output pulse for two input pulses. This fact is fundamental to the operation of counters and shift registers.

Let us first of all remind ourselves of the binary counting sequence. A 4-bit sequence is shown in Figure 7.1, with A being the least significant bit and D the most significant. The count goes from 0 to 15 and there are 16 lines in the table.

	Binary			
D	*C*	*B*	*A*	*Decimal*
2^3(8s)	2^2(4s)	2^1(2s)	2^0(1s)	
0	0	0	0	0
0	0	0	1	1
0	0	1	0	2
0	0	1	1	3
0	1	0	0	4
0	1	0	1	5
0	1	1	0	6
0	1	1	1	7
1	0	0	0	8
1	0	0	1	9
1	0	1	0	10
1	0	1	1	11
1	1	0	0	12
1	1	0	1	13
1	1	1	0	14
1	1	1	1	15

Figure 7.1 *The counting sequence table*

Counters

The 3-bit asynchronous counter

Figure 7.2 shows three JK flipflops connected in series. This example assumes negative edge triggering, which means that all changes take place on the negative or trailing edge of the clock pulse.

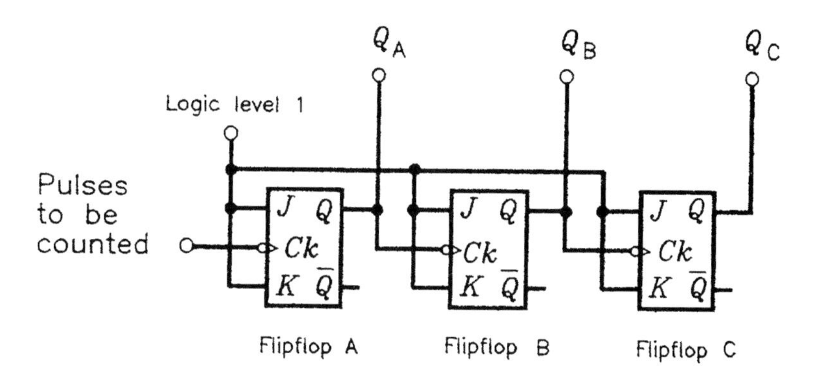

Figure 7.2 *The 3-bit asynchronous counter using JK flipflops*

Important points

- The pulses to be counted are fed into the clock input (Ck) of the first flipflop (A).
- The Q **output** from a particular flipflop is used as the **clock input** for the next.
- The J and K inputs of all flipflops are connected together to logic 1, i.e. +5 V. The flipflops are thus operating in their toggling mode. For a reminder of toggling, see Chapter 6.
- Flipflop A (the input), providing the **least significant binary digit**, is on the left-hand side of the diagram. This is a matter of convention, since inputs are traditionally shown on the left-hand and outputs on the right-hand side of diagrams.

  There should be no confusion with the counting sequence table (which shows A on the right-hand side) as long as it is remembered that A is the least significant bit.

The action of the counter is illustrated by the waveform diagrams in Figure 7.3. Assume that all flipflop outputs are initially at logic 0 and that all changes take place on the trailing, or negative, edge of the input (clock) pulse.

The output Q_A will go through one complete cycle, or pulse, from time $t = 0$ to time $t = t_1$, for two cycles (pulses) of the input. Thus flipflop A divides by two! Notice

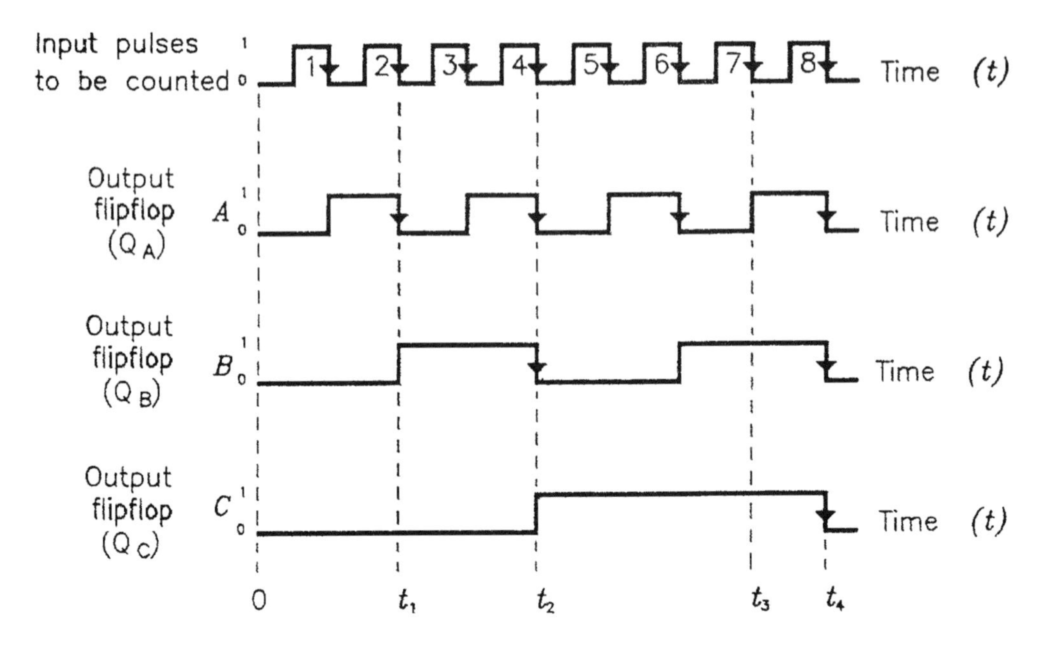

Figure 7.3 *The 3-bit asynchronous counter: waveform diagrams*

that the changes in level of output Q_A occur only when the input to flipflop A is negative-going. Similarly, for two input pulses to flipflop B (and hence four input pulses to flipflop A) there is one pulse at the output Q_B. This process is continuous and, for example, after eight input clock pulses to flipflop A (time $t = t_4$) there will be four pulses at Q_A, two at Q_B and one at Q_C, with each flipflop thus dividing by two. Figure 7.4 is a tabular version of the waveforms of Figure 7.3 taken at three instants in time. Remember that Q_A, Q_B and Q_C represent 2^0 (=1), 2^1 (2) and 2^2 (4) respectively.

Time (t)	Number of input pulses to flipflop A	Logic level: input pulse	Logic level: output Q_C (2^2)	Logic level: output Q_B (2^1)	Logic level: output Q_A (2^0)	output count ($2^2 + 2^1 + 2^0$)
t_1	2	0	0	1	0	0 + 2 + 0 = 2
t_2	4	0	1	0	0	4 + 0 + 0 = 4
t_3	7	0	1	1	1	4 + 2 + 1 = 7

Figure 7.4 *The 3-bit asynchronous counter: looking at various points on the waveform diagram of Figure 7.3*

Referring back to Figure 7.3, at time $t = t_4$, we see that all output logic levels are 0. If there were a fourth bit D we would notice at this point that its logic level was 1, thus giving the required output count of 8.

Important points

- The 3-bit counter will count up to 7, and will reset automatically on the 8th pulse. It is known as a **modulo 8 counter**, the term 'modulo' being derived from modulus, meaning magnitude.
- Another way of looking at modulus is to regard it as the number of different output states the counter must go through before resetting to zero.
- In general, an n-bit (modulo n) counter, will

  divide by n
  count up to the number 2^{n-1}
  reset on the $2^{n\text{th}}$ pulse

- Look again at the time interval $t = 0$ to $t = t_4$ on Figure 7.3. Notice that the action of each flipflop following the input of clock pulse 1 is staggered. This is because each flipflop after the first receives its input from the preceding one. Thus, only the first flipflop is clocked by an external pulse. The term used to describe the operation of this type of flipflop is **asynchronous** which means 'not synchronous'.

  The application of the input pulse takes time to ripple through the chain of flipflops, giving the name **ripple counter** to this type of device.
- Since each flipflop will have a certain propagation delay (15 ns is quoted for the 74LS76 ic), the time for the operation to be completed for the 3-bit counter above is of the order of 45 ns. This illustrates the main disadvantage of the asynchronous counter, and limits the maximum clock frequency that can be used. In addition, if the output is being decoded there can be problems with unwanted spikes or glitches.
- Taking the Q output as the means of clocking the following flipflop gives a **counting up sequence**, with the counter going from 0 to 7 and resetting on 8.

  Remembering that the $\overline{Q}$ outputs are the inverse of Q, then taking these $\overline{Q}$ outputs instead will enable the counter to **count down**. Note that the count output is still taken from Q.

Practical Exercise 7.1

To investigate the 3-bit asynchronous binary counter
For this exercise you will need the following components and equipment:

2 – 74LS76 ic (dual JK flipflop with preset and clear active low, and negative edge triggering)
3 – LED (5 mm) and resistor (270 Ω)

Continued on p. 132

Practical Exercise 7.1 (*Continued*)

1 – +5 V DC power supply
1 – pulse generator (1 Hz and 1 kHz) (see Figure 1.5)
1 – double beam cathode ray oscilloscope

Procedure

1 Connect up the circuit shown in Figure 7.5. The pin connection diagram for the 7476 ic is given in Figure 6.33.

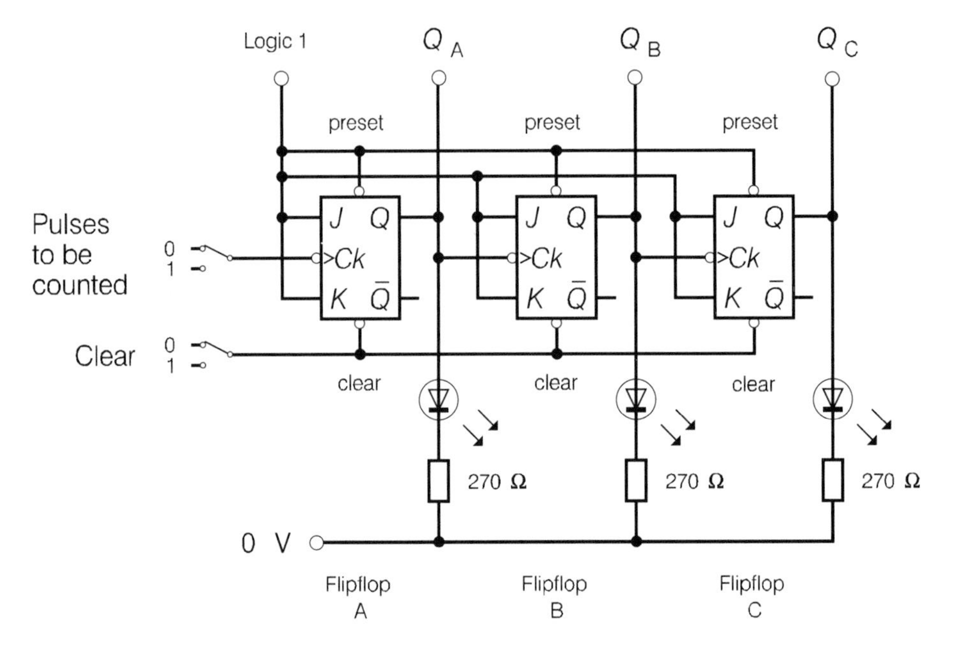

Figure 7.5 *The 3-bit asynchronous binary counter, counting up: circuit for Practical Exercise 7.1*

2 Connect all preset inputs to logic 1.
3 Initially, connect all clear inputs to logic 0 (thus ensuring that all outputs are reset to logic 0) and then leave them connected to logic 1.
4 Apply a series of +5 V pulses to the clock input of flipflop A. Alternatively, use the pulse generator to provide a low frequency (1 Hz) output.
5 The counter should go through the counting sequence from 0 to 7, resetting to 0 on the 8th pulse. This is counting up.

Note that the output can be reset to logic 0 at any time in the sequence, by returning the clear inputs to logic 0.

6 With the higher-frequency clock pulse (1 kHz), use the cathode ray oscilloscope to inspect the output (Q_A to Q_C) waveforms, relating them separately to the clock waveform.

Draw the four waveforms one underneath the other and make sure that they are time-related (as in Figure 7.3).

7 Change the output connections (see Figure 7.6) to enable the counter to count down. Remember to clear the Q outputs to logic 0 if necessary.

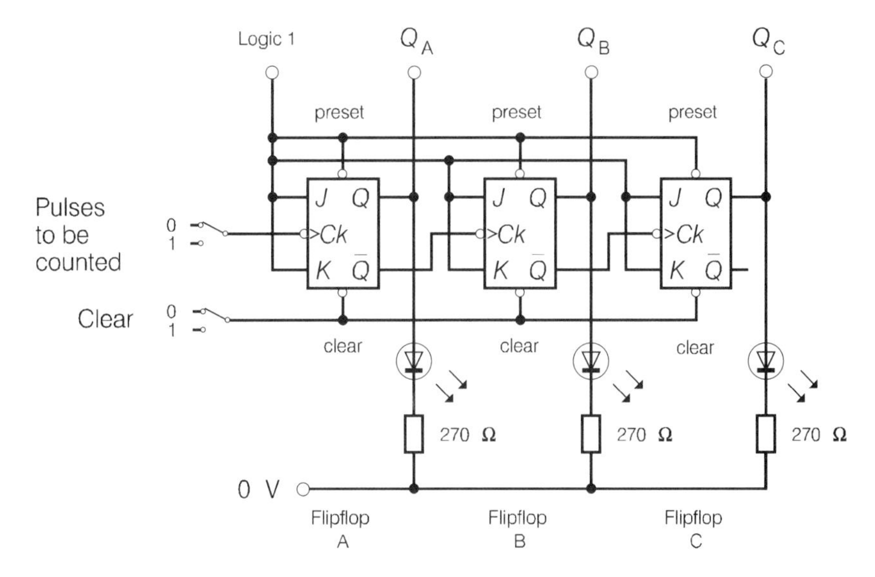

Figure 7.6 *The 3-bit asynchronous binary counter, counting down: circuit for Practical Exercise 7.1*

8 Check that the counting sequence is now from 7 to 0, resetting to 7 on the 8th pulse.

9 Use the generator and oscilloscope as before. Notice for counting down, that while the output Q_A changes on the negative edge of the clock pulse, as previously, the outputs Q_B and Q_C change when the previous flipflop Q output is positive-going.

Draw the four waveforms one underneath the other and again make sure that they are time-related.

Conclusions

From the waveforms drawn in procedure 9 above, explain why the leading edge of the waveforms for Q_B and Q_C occur on the leading edge of those for Q_A and Q_B respectively.

Hint: It may be helpful to your understanding to include the waveforms of $\overline{Q}_A$ and $\overline{Q}_B$.

Stopping the count

It is sometimes useful and necessary to count to a number occurring before the device automatically resets itself. For example, suppose we need to count up to 10, that is, to rest on the 10th pulse, which then becomes the count of zero.

Look again at the counting sequence in Figure 7.1. The count of 10 occurs when $A = 0$, $B = 1$, $C = 0$ and $D = 1$ and it now becomes necessary to reset the flipflop, using in this example the logic 1 outputs from Q_B and Q_D.

If the flipflops used are active-low, then a logic 0 at the reset input of these flipflops will cause the counter to be reset. The question is, what gate will provide a logic 0 output when all its inputs are logic 1? (The answer is, a NAND gate.)

The next practical exercise will use the 7493 ic, firstly as a 4-bit counter to count up to 15, and then by using the reset inputs, as a decade (count-to-10) counter.

Practical Exercise 7.2

To investigate the reset action of the 4-bit asynchronous binary counter
For this exercise you will need the following components and equipment:

1 – 74LS93 ic (4-bit binary counter)
4 – LED (5 mm) and resistor (270 Ω)
1 – +5 V DC power supply
1 – pulse generator (1 Hz and 1 kHz) (see Figure 1.5)
1 – double beam cathode ray oscilloscope

The 7493 ic is a 4-bit device, arranged as a divide-by-2 (using flipflop A only) and divide-by-8 (flipflops B to D), with separate inputs and outputs. Thus, by using all four flipflops it will behave as a divide-by-16 (modulo 16) counter. The use of the reset inputs in conjunction with the internal NAND gate will enable the count to be stopped (reset) at certain points in the sequence. The reset inputs labelled R_{01} and R_{02} require a logic 1 and will reset the counter to 0.

Procedure

1. The pin connections for the 7493 ic are given in Figure 7.7. Note that pin 12 must be externally connected to pin 1.
2. Make up the 4-bit counter using Figure 7.5 as a guide. Note that the J and K inputs are internally connected to the +5 V supply.
3. Connect R_{01} and R_{02} to logic 0. Note that the 7493 ic has no preset inputs and no $\overline{Q}$ outputs.
4. Apply a series of +5 V pulses to the clock input of flipflop A as before (or use the generator).
5. The counter should go through the counting sequence from 0 to 15, resetting on the 16th pulse.

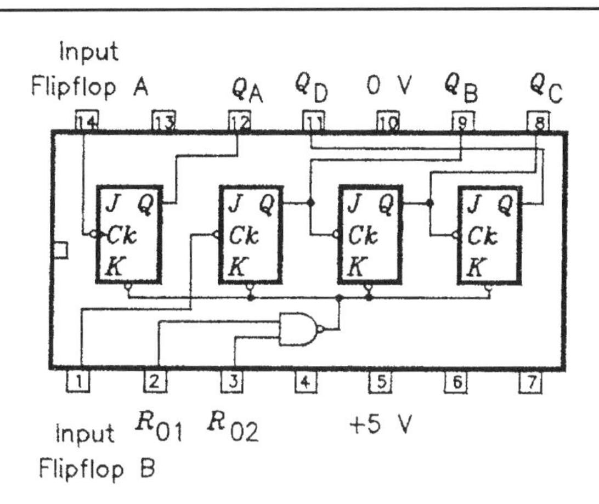

Figure 7.7 *The 7493 binary counter: pin connections*

6 With the higher frequency pulse (1 kHz) as the clock input, as in the previous exercise, have a look at the waveforms at each of the outputs Q_A to Q_D and relate them separately to the clock input waveform.
7 Draw the waveform diagram, similar to Figure 7.3, but appropriate to the 4-bit counter.
8 Return to a low frequency clock pulse (or the +5 V supply) and connect the reset inputs R_{01} and R_{02} to outputs Q_B and Q_D respectively.
9 The count should now be from 0 to 9, resetting on the 10th pulse. This is now the **decade counter**.

Questions

7.1 Draw the circuit diagram of a 4-bit ripple counter using JK flipflops. Explain how the counter arrives at the count of 1011 (decimal 11). What is the maximum count available?
7.2 Show the additional circuitry that will convert the counter in question 1 to a (a) modulo 6, (b) modulo 12 counter. Draw a set of waveform (timing) diagrams to illustrate the modulo 6 counter.
7.3 A suitable purpose-built decade counter is the 7490 ic, which will be used in a later practical exercise. Look at the pin connection diagram in Figure 10.14 and explain the purpose of the R_0 and R_9 pairs of inputs.

Figure 7.7 showed the arrangement using a 2-input NAND gate, which enabled the 7493 4-bit binary counter to be reset on the count of 10. Suppose that it was necessary to reset on, say, the 7th pulse, where the binary count was $ABCD = 0111$. A 3-input NAND gate would be necessary here, in order to use all the 1's.

If the requirement was to be able to reset on any count between 1 and 15, a 4-input NAND gate would be necessary, since output D (the 4th) is logic 1 from the count of 8.

Remembering that no inputs to the gate must be left unconnected (that is, floating), then for the 4-bit decade counter (with output $ABCD = 0101$), the inputs for $B = D = 1$ must come from Q_B and Q_D respectively and those for $A = C = 0$ (or NOT 1) from $\overline{Q}_A$ and $\overline{Q}_C$ respectively.

The 3-bit synchronous counter

The disadvantage of asynchronous counters, namely the build-up of time delay which can be longer than the time between input pulses, has already been mentioned. The synchronous counter, where all flipflops are under the control of the clock, is shown in Figure 7.8. The flipflops are **clocked together** and there is only one delay time for this

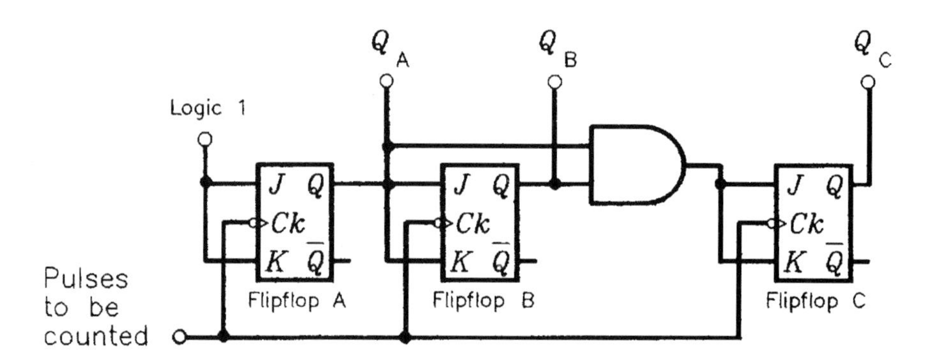

Figure 7.8 *The 3-bit synchronous counter*

Clock pulse	*Flipflop A* JK inputs	Q_A output	*Flipflop B* JK inputs	Q_B output	*AND gate A* inputs Q_A Q_B	output	*Flipflop C* JK inputs	Q_C output	*Binary Count* Q_C Q_B Q_A
0	1	0	0	0	0 0	0	0	0	0 0 0
1	1	0 to 1 (toggles)	0 to 1	0	1 0	0	0	0	0 0 1
2	1	1 to 0 (toggles)	1 to 0	0 to 1 (toggles)	0 1	0	0	0	0 1 0
3	1	0 to 1 (toggles)	0 to 1	1	1 1	1	0 to 1	0	0 1 1
4	1	1 to 0 (toggles)	1 to 0	1 to 0 (toggles)	0 0	0	1 to 0	0 to 1 (toggles)	1 0 0
5	1	0 to 1 (toggles)	0 to 1	0	1 0	0	0	1	1 0 1
6	1	1 to 0 (toggles)	1 to 0	0 to 1 (toggles)	0 1	0	0	1	1 1 0
7	1	0 to 1 (toggles)	0 to 1	1	1 1	1	0 to 1	1	1 1 1
8	1	1 to 0 (toggles)	1 to 0	1 to 0 (toggles)	0 0	0	1 to 0	1 to 0 (toggles)	0 0 0

Figure 7.9 *The 3-bit synchronous counter: counting sequence*

particular action, plus whatever there may be for the control (AND) gate(s). Thus the effect of this synchronous action is a very much reduced propagation delay, allowing a typical clock input frequency of 32 MHz (twice that of the asynchronous equivalent).

The counting sequence is shown in Figure 7.9.

Practical Exercise 7.3

To investigate the action of the 3-bit synchronous counter
For this exercise you will need the following components and equipment:

2 – 74LS76 ic (dual JK flipflop)
1 – 74LS08 ic (quad 2-input AND)
3 – LED red (5 mm) and resistor (270 Ω)
1 – LED green (5 mm) and resistor (270 Ω)
1 – +5 V DC power supply
1 – pulse generator (1 Hz) (see Figure 1.5)
1 – double beam cathode ray oscilloscope

Procedure

1 Make up the circuit of Figure 7.10. See Figure 6.28 for the 7476 pin connections and Figure 2.13(a) for the 7408.

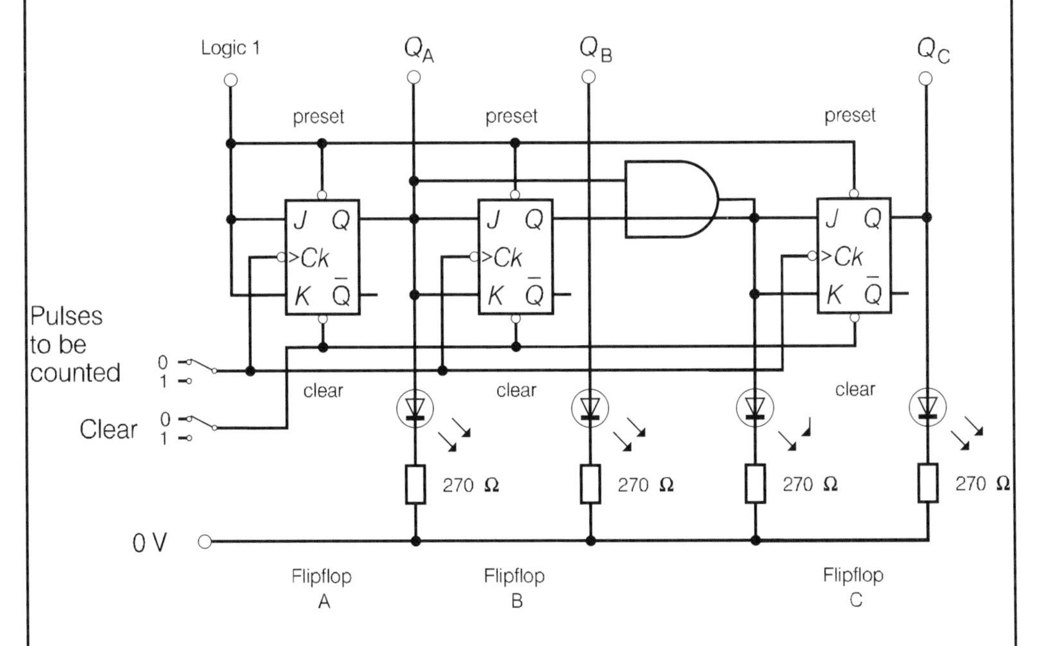

Figure 7.10 *The 3-bit synchronous counter: circuit for Practical Exercise 7.3*

Continued on p. 138

Practical Exercise 7.3 *(Continued)*

2 Connect the clear inputs initially to logic 0 (to reset the counter to 000) and then permanently to logic 1.
3 The clock input can again be obtained from either the +5 V supply or the generator.
4 Work through the sequence of Figure 7.9 and confirm that the counter counts from 0 to 7 and resets on the 8th pulse. In particular, for each of the eight clock pulses, check the state of the four LEDs with the expected result.

The 4-bit synchronous counter

The diagram for the 4-bit counter is shown in Figure 7.11. The reader may like to produce the counting sequence table, on the style of Figure 7.9.

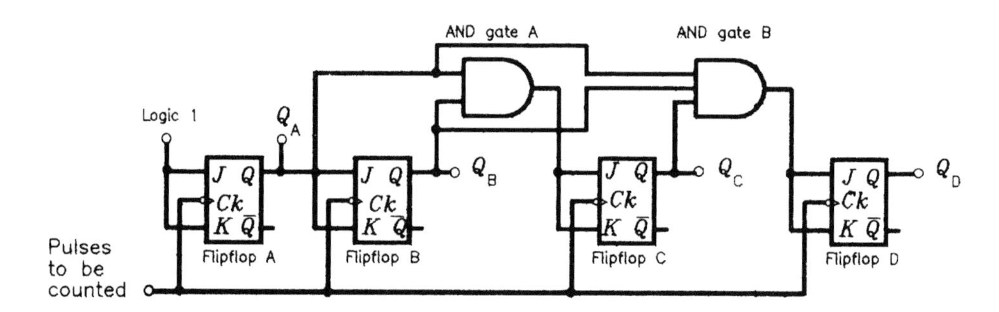

Figure 7.11 *The 4-bit synchronous counter*

This counter could be made up with individual components as in Practical Exercise 7.3. However, since synchronous counters are available in ic form, this will be the basis of the next two exercises.

The ic's available within the TTL family are as follows:

74160	4-bit decade counter with asynchronous reset
74161	4-bit binary counter with asynchronous reset
74162	4-bit decade counter with synchronous reset
74163	4-bit binary counter with synchronous reset
74169	4-bit binary up–down counter
74190	4-bit BCD up–down counter
74191	4-bit binary up–down counter
74192	4-bit BCD up–down counter
74193	4-bit binary up–down counter

Remember that:

(a) the asynchronous reset will override all other inputs;
(b) the synchronous reset will operate with the clock pulse, which for this type of counter is usually the positive edge;
(c) BCD = binary coded decimal.

Practical Exercise 7.4

To investigate the 4-bit synchronous up counter
For this exercise you will need the following components and equipment:

1 – 74LS161 ic (4-bit synchronous binary counter)
1 – 74LS00 ic (quad 2-input NAND)
4 – LED red (5 mm) and resistor (270 Ω), for the *Q* outputs
1 – LED green (5 mm) and resistor (270 Ω), for the carry output
1 – +5 V DC power supply

Procedure (a): to count to decimal 16

1 Make up a suitable circuit. The pin connection diagram for the 74161 ic is shown in Figure 7.12.

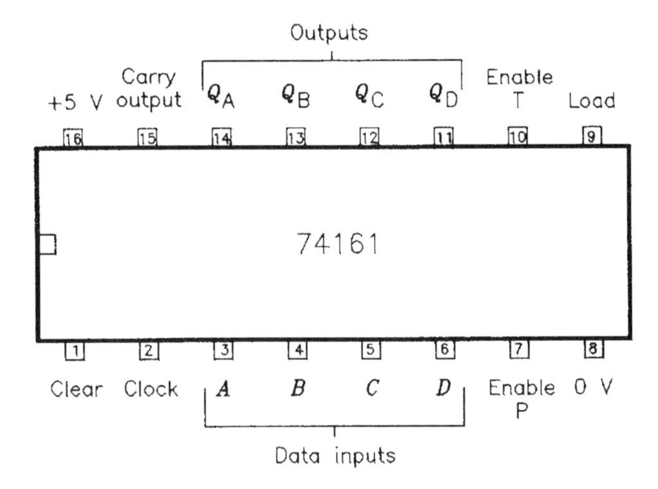

Figure 7.12 *The 74161 synchronous binary counter: pin connections*

2 Make the following settings:
Load input – logic 1
Enable inputs (P&T) – logic 1

Continued on p. 140

Practical Exercise 7.4 *(Continued)*

Data inputs – either logic 1 or logic 0 (to leave them floating can cause erratic behaviour of the counter)
Clear – logic 0 (to reset) then to logic 1

3 Apply a series of +5 V pulses to the clock input. The counter should count from *ABCD* = 0000 to *ABCD* = 1111 + carry 1 and then reset to *ABCD* = 0000.
Note. Counting action takes place on the positive edge of the clock pulse.

Procedure (b): to stop the count before decimal 15

1 Modify the circuit to include the NAND gate whose pin connections are given in Figure 2.13(b).
The action is now identical to that met in Practical Exercise 7.2, in that the counter will reset when the inputs to the NAND gate are both logic 1.
Suppose that we want the counter to count to decimal 10 which means resetting when outputs *ABCD* = 0101.
2 Make the following settings:

Load input – logic 1
Enable inputs (P&T) – logic 1
Data inputs *ABCD* – either logic 1 or logic 0 (as above)

3 Apply a series of +5 V pulses to the clock input. The counter should count from *ABCD* = 0000 and reset at *ABCD* = 0101.
4 Repeat Procedure 2 with a reset value of your choice.

Procedure (c): to preset the counter

1 Return to the circuit of Figure 7.12. Suppose that we want the counter to start counting at *ABCD* = 0101.
2 Make the following settings:

Data inputs *ABCD* – 0101 (= preset value)
Load input – logic 0 (this will disable the counter)
Enable inputs (P&T) – logic 1

3 Apply one clock pulse. The output should go to its preset value (on the positive edge of the pulse).
4 Now set the load input to logic 1 and apply further clock pulses. The count should proceed from *ABCD* = 0101 to *ABCD* = 1111 + carry 1 and then reset to *ABCD* = 0000.
5 Reload the counter to a preset value of your choice.

Practical Exercise 7.5

To investigate the 4-bit synchronous up–down counter
For this exercise you will need the following components and equipment:

1 – 74LS192 ic (4-bit synchronous up–down BCD counter)
4 – LED red (5 mm) and resistor (270 Ω), for the *Q* outputs
1 – LED green (5 mm) and resistor (270 Ω), for the carry output
1 – LED yellow (5 mm) and resistor (270 Ω), for the borrow output
1 – +5 V DC power supply

The 74192 4-bit synchronous counter will count up to and down from decimal 10. In addition the start of each count can be preset. A carry or borrow output is available.
The pin connection diagram is shown in Figure 7.13

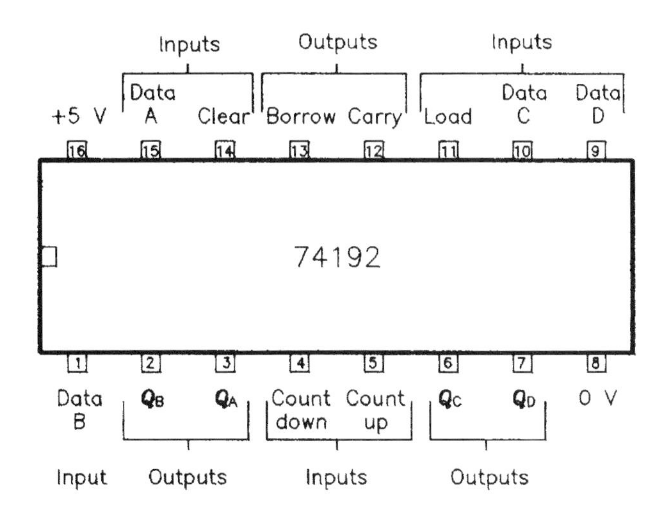

Figure 7.13 *The 74192 synchronous up–down BCD counter: pin connections*

Procedure (a): to count up

1 Make the following settings:

Load input – logic 1
Count down input – logic 1
Data inputs – logic 1 or logic 0
Clear input – logic 1 then logic 0

Continued on p. 142

Practical Exercise 7.5 *(Continued)*

2 Apply a series of +5 V pulses to the count up input. The counter should count from *ABCD* = 0000 to *ABCD* = 1001 then reset to *ABCD* = 0000 carry 1.

Note. The borrow output will be at logic 1 all the time. The carry output will be at logic 1 all the time except when going to logic 0 on the negative edge of the count *ABCD* = 1001, then returning to logic 1 at *ABCD* = 0000.

Procedure (b): to count down

1 Make the following settings:

Load input – logic 1
Count up input – logic 1
Data inputs – logic 1 or logic 0
Clear – logic 1 then logic 0.

2 Apply a series of +5 V pulses to the count down input. The counter should count from *ABCD* = 1001 to *ABCD* = 0000 then reset to *ABCD* = 1001 borrow 1.

Note. The carry output will be at logic 1 all the time. The borrow output will be at logic 1 all the time except when going to logic 0 on the negative edge of the count *ABCD* = 0000, then returning to logic 1 at *ABCD* = 1001.

Procedure (c): to count up from a preset value (say, ABCD = 1110)

1 Make the following settings:

Load input – logic 0 (to disable counter)
Count down input – logic 1
Data inputs *ABCD* – 1110
Clear input to logic 1 then logic 0 (the outputs will take the level *ABCD* = 1110 without the need for a count up pulse)
Load input – logic 1.

2 Apply a series of +5 V pulses to the count up input. The counter should count from *ABCD* = 1110 to *ABCD* = 0000 carry 1, then continuing from *ABCD* = 1000 carry 1.

Note. A carry output is present all the time except when going to zero on the negative edge of the count *ABCD* = 1001, then returning for *ABCD* = 0000.

Procedure (d): to count down from a preset value (say, ABCD = 1110)

1 Make the following settings:

Load input – logic 1
Count up input – logic 1
Data inputs *ABCD* – 1110.

2 Apply a series of +5 V pulses to the count down input. The counter should count from $ABCD = 1110$ to $ABCD = 0000$ borrow 1, then continuing from $ABCD = 1001$ borrow 1.

Note. A borrow output is present all the time except when going to zero on the negative edge of the count $ABCD = 0000$, then returning for $ABCD = 1001$.

Summary of the necessary procedures

To count up/down 'normally'

(a) Load input – logic 1
(b) Count down/up input – logic 1
(c) Data inputs – either logic 1 or logic 0
(d) Clear input – logic 1 (to reset) then logic 0
(e) Apply +5 V pulse to count up/down input

Note. Borrow output always at logic 1. Carry input at logic 1 except on negative edge of pulse immediately before the end of the natural count.

To count up/down from a preset value

(a) Load input – logic 0 (disables counter for loading)
(b) Count down/up input – logic 1
(c) Data inputs to preset level. The outputs will take the preset level without the need for a count up/down pulse
(d) Load input – logic 1
(e) +5 V pulses to up/down input

Note. Carry output always at logic 1. Borrow input at logic 1 except on negative edge of pulse immediately before the end of the natural count.

Question

7.4 Explain the operational difference between asynchronous and synchronous counters and state any advantage in using the latter.

Shift registers

Take your calculator for a moment and enter in a series of numbers. Notice that, as you enter successive numbers from the keyboard, the numbers in the display window shift to the left. You will also see that as you enter each number, the previous number is held (memorized) in the display.

> **Important points**
>
> - A shift register is a temporary store of binary data.
> - It consists of a number of flipflops, each of which will store one bit of data.
> - The data can then be shifted (moved) at a later time, for further processing.
> - This shift can be either to the left or to the right within the register, in order to perform a particular arithmetic operation.

The serial load shift register

The circuit of a 4-bit serial load shift register, using D flipflops, is shown in Figure 7.14.

Firstly, a look at the word **serial**: this is another name for series – notice that the data enters the register at the input of flipflop A and can only do so one bit at a time. This is the serial loading referred to above. The outputs are taken from each of the flipflops (Q_A to Q_D) as shown and are parallel outputs. The complete name for this type of shift register is **serial in–parallel out**, usually abbreviated to SIPO.

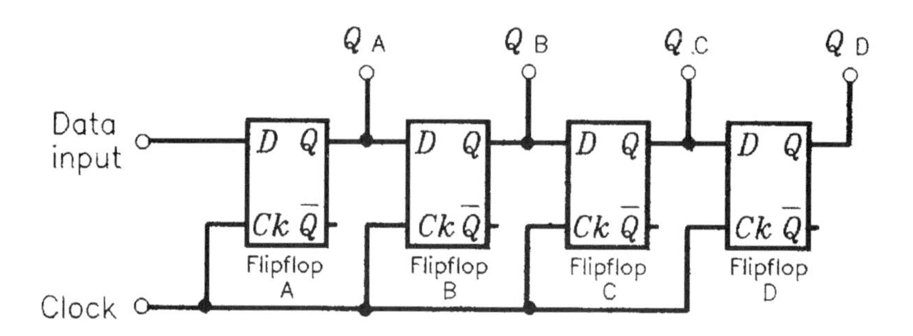

Figure 7.14 *The 4-bit serial load shift register using D flipflops*

Clock pulse	Data *D*	Q_A	Q_B	Q_C	Q_D	Action
0	0	0	0	0	0	All flip–flops reset to logic 0
0	1	0	0	0	0	Outputs remain at logic 0
1	1	1	0	0	0	Data enters register at output Q_A
2	0	0	1	0	0	Data moves to output Q_B
3	0	0	0	1	0	Data moves to output Q_C
4	0	0	0	0	1	Data moves to output Q_D
5	0	0	0	0	0	Data leaves register

Figure 7.15 *The shift register: table showing movement of data through the register*

Now let us look at the action of this register. Remember that the Q output of the D-type flipflop will take the same logic level as the D input only when the clock pulse is applied: that is, on either the leading or the trailing edge of the clock pulse.

Assume that all flipflops are reset (cleared), such that Q_A to Q_D outputs are at logic 0. Now, with the D (data) input of flipflop A at logic 1, nothing happens until the first clock pulse comes along, at which time the output Q_A will go to the same logic level as its D input, which in this case is logic 1. At this point all the other outputs will remain at logic 0 since their D inputs are at logic 0. No further changes can take place until the next clock pulse.

On the second clock pulse, again, the Q outputs of all flipflops will take the same level as their D input. This means, assuming that the data input to flipflop A is back to logic 0, that the output Q_A will go to logic 0, the output Q_B will go to logic 1, and the other outputs Q_C and Q_D will remain at logic 0. Thus the original data input of logic 1 has been **shifted** from the output of flipflop A to that of flipflop B following the application of the second clock pulse.

Keeping the data input to flipflop A at logic 0, and applying two more (the third and fourth) clock pulses, it will be apparent that the data will shift to the output of flipflop D. It has taken a total of four clock pulses to shift the data from the **input** of flipflop A to the **output** of flipflop D.

One more (the fifth) clock pulse will see all outputs at logic 0. The data has moved through the register, which is now empty.

When data moves from the least significant stage (flipflop A) to the most significant (in this case, flipflop D), the register is known as a **shift-right register**. The truth table for this register is given in Figure 7.15. The waveform diagrams, showing the operation for both positive and negative edge triggering, are given in Figure 7.16.

Practical Exercise 7.6

To investigate the 4-bit serial load shift register using D flipflops
For this exercise you will need the following components and equipment:

2 – 74LS74 ic (dual D flipflop)
4 – LED (5 mm) and resistor (270 Ω)
1 – +5 V DC power supply

The circuit is shown in Figure 7.17 and the pin connections for the 7474 ic in Figure 6.23.

Procedure (a)

1. Clear the register by setting all the clear inputs to logic 0, after which they should be returned to logic 1 (otherwise the register will remain permanently cleared!). At this point, all outputs should be at logic 0 and the register can be regarded as a serial in–parallel out type.

Continued on p. 146

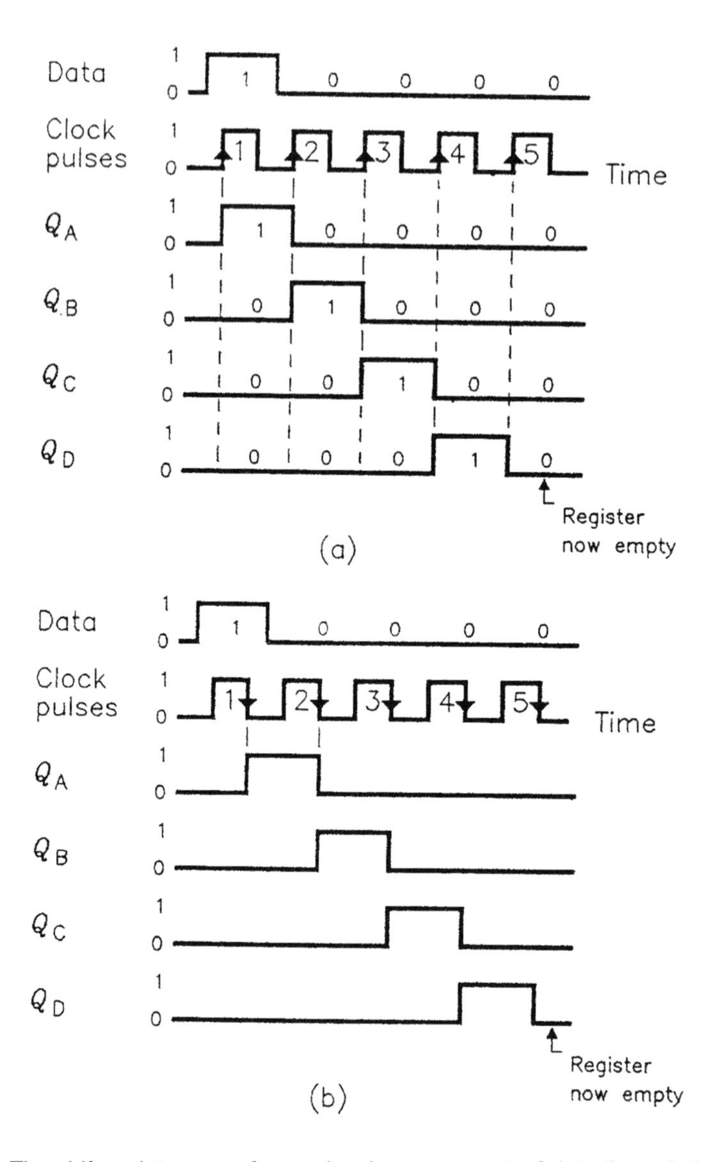

Figure 7.16 *The shift register: waveforms showing movement of data through the register*

Practical Exercise 7.6 (*Continued*)

2 The preset inputs (which are **active low**) can be left unconnected, in which case their voltage levels will tend to go towards logic 1, but may not reach its threshold value. It would be wise therefore to connect all these inputs to logic 1.
3 The clock input should be a series of one-off pulses (+5 V).
4 With the clock input at logic 0, set input *D* to logic 1. All outputs **should still be at logic 0**.

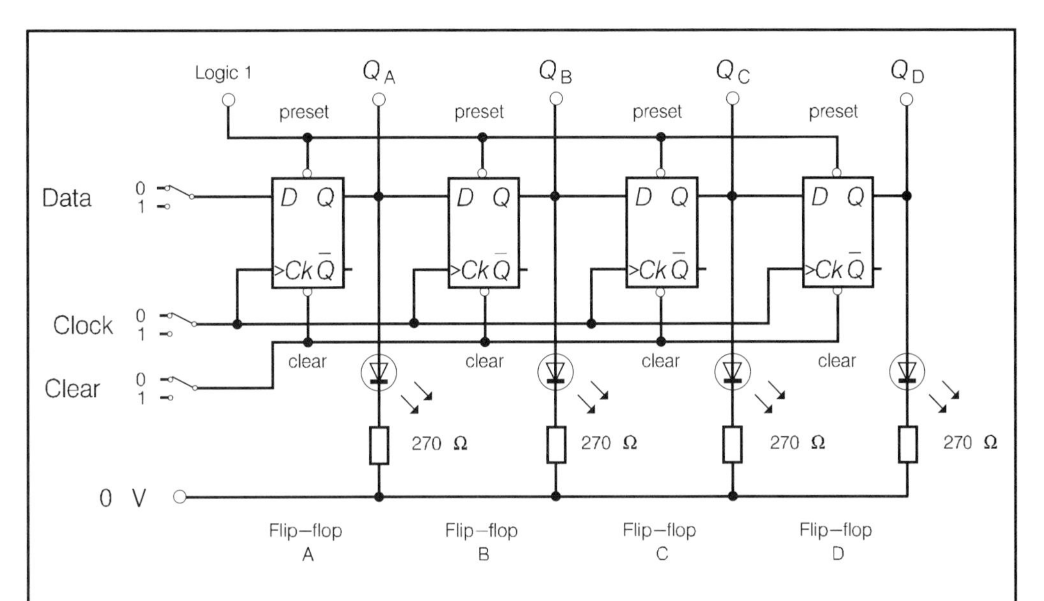

Figure 7.17 *The 4-bit serial load shift register: circuit for Practical Exercise 7.6*

5 Set the clock input to logic 1 and note the state of the outputs. Q_A should go to logic 1 when the clock goes from logic 0 to logic 1, which is **positive edge** triggering.

6 Return the D input to logic 0 and apply a succession of clock pulses, noting the state of the outputs. You should be able to verify the truth table shown previously in Figure 7.15. The movement of data from the D input to the final output Q_D is **serial in–serial out operation** (SISO).

7 Now to investigate the effect of having a preset output: clear the register and set the preset input of flipflop B to logic 0 (leaving all the others at logic 1). Note that output Q_B is now logic 1 and that setting the clear inputs to logic 0 has no effect on this output.

8 Perform the sequences shown as (a), (b) and (c) in Figure 7.18. Copy and complete the tables.

9 Return the preset of flipflop B to logic 1.

10 Finally, check the operation of this register as a **serial in–parallel out** type, by loading into the register the data $ABCD = 0101$ and noting the number of clock pulses necessary for this.

Note. In this example, loading means entering the data one bit at a time at the D input and shifting it to the right as necessary.

Conclusions

1 State the answer obtained in procedure 10.

2 How many clock pulses were necessary in procedures 5 and 6 for the single bit of data to move from the D input to the Q_D output?

Continued on p. 148

Practical Exercise 7.6 (*Continued*)

(a)

Input D	Clock pulse	Q_A	Q_B	Q_C	Q_D
0	0	0	1	0	0
0	1				
0	2				
0	3				

(b)

Input D	Clock pulse	Q_A	Q_B	Q_C	Q_D
1	0	0	1	0	0
1	1				
1	2				
1	3				

(c)

Input D	Clock pulse	Q_A	Q_B	Q_C	Q_D
1	0	0	1	0	0
1	1				
0	2				
0	3				

Figure 7.18 *Examples for Practical Exercise 7.6*

Procedure (b)

You will have noticed, from procedure (a)6, that following the fifth clock pulse, the register was empty with all Q outputs showing logic 0.

1. Connect the output of flipflop D (i.e. output Q_D) to the D input of flipflop A. This creates a feedback link from output to input.
2. Repeat procedures (a) 1 to 5.
3. Return the D input to logic 0 and apply a succession of clock pulses, noting the state of the Q outputs. Hence comment on the result of having the feedback link.

4 Clear the register and make the preset input of flipflop B logic 0. This should make output Q_A logic 1, giving a display of 1000.

5 Apply a series of clock pulses and note how many are needed for the 1000 display to next appear.

Important points

- The feedback link causes the data to **recirculate** in the register. It is then, not surprisingly, known as a recirculating shift register.
- A shift register whose input is obtained from its output, as above, is also known as a **ring counter**. See Figure 7.19.

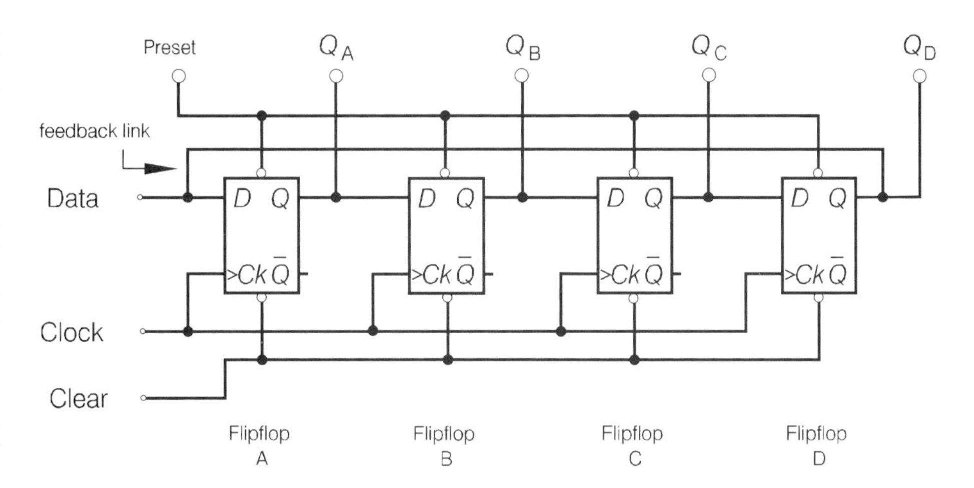

Figure 7.19 *The 4-bit serial load recirculating shift register: circuit for Practical Exercise 7.6(b). (The ring counter)*

- This ring counter is decoded as follows:

  With Q_A at logic 1, the count is zero
  With Q_B at logic 1, the count is one
  With Q_C at logic 1, the count is two
  With Q_D at logic 1, the count is three

  A count of ten would thus require 10 flipflop stages, which is not a very economical arrangement compared with the 7490 decade counter.

An alternative circuit for the serial load shift register uses JK flipflops wired as D-types, and this is shown in Figure 7.20. The use of the inverter gate in the circuit of Figure 7.20 ensures that inputs J and K are always complementary.

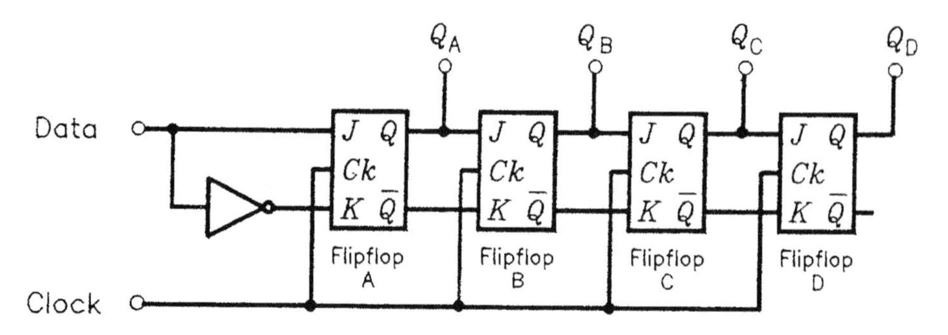

Figure 7.20 *The 4-bit serial load shift register using JK flipflops*

The important characteristic of the serial shift register is that data is entered in or taken out in sequence, that is, one bit at a time. The speed at which data can be shifted is thus an important factor. While this is a basic disadvantage where it is necessary to transmit information quickly, the built-in delay provides one application for this type of shift register. For example, the total time delay introduced by a four-stage serial in–serial out register, will be equal to the time for four clock pulses.

PISO and PIPO shift registers

There are two other types of shift register:

(a) parallel in–series out (PISO)
(b) parallel in–parallel out (PIPO)

The feature of the parallel type is that data is fed into and taken out from all flipflops at the same time. The circuit diagram of a 4-bit parallel load shift register is shown in Figure 7.21. (Practical Exercise 7.7).

Practical Exercise 7.7

To investigate the 4-bit parallel load shift register
For this exercise you will need the following components and equipment:

2 – 74LS76 ic (dual JK flipflop)
4 – LED (5 mm) and resistor (270 Ω)
1 – +5 V DC power supply

Procedure

1 Connect up the circuit shown in Figure 7.21. Refer to Figure 6.33 for the pin connections of the 7476 ic.

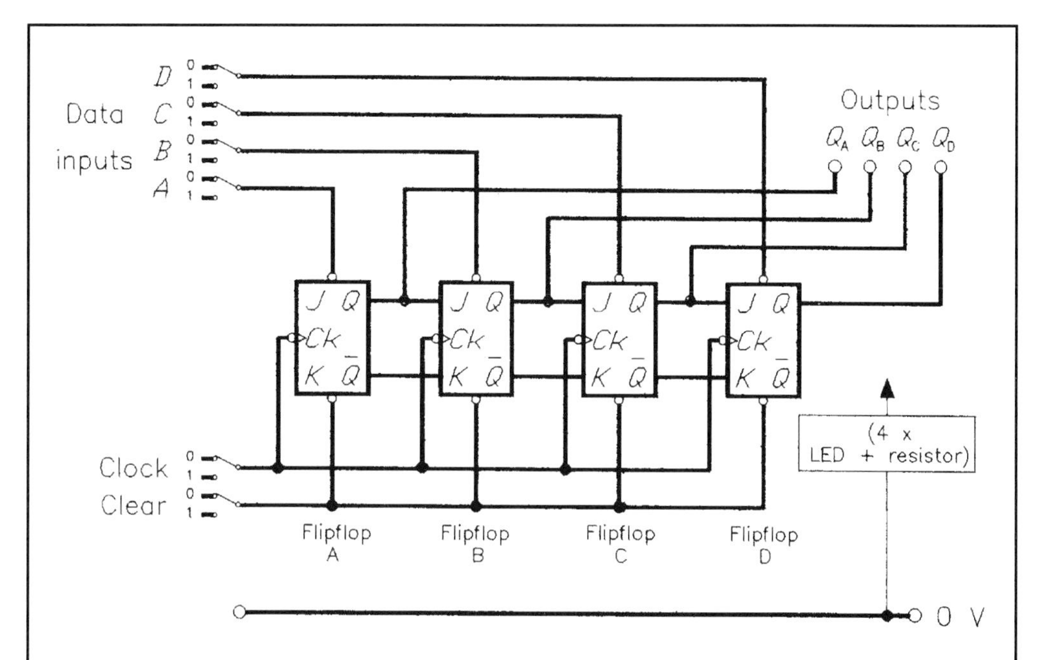

Figure 7.21 *The 4-bit parallel load shift register using JK flipflops: circuit for Practical Exercise 7.7*

Connect the LED indicators and their series resistors to the Q outputs as in previous exercises.

2 With all presets at logic 1, clear the register by connecting the clear inputs to logic 0. This is the first line of the truth table given in Figure 7.22. The clear input should then be returned to logic 1.

Clear	Parallel load A B C D	Clock pulse	$Q_A Q_B Q_C Q_D$	Comments
0	1 1 1 1	0	0 0 0 0	Clear is 'active low'
1	0 1 1 1	0	1 0 0 0	Loading is 'active low'
1	1 1 1 1	1	0 1 0 0	Data shifts to the right
1	1 1 1 1	2	0 0 1 0	
1	1 1 1 1	3	0 0 0 1	
1	1 1 1 1	4	0 0 0 0	

Figure 7.22 *The 4-bit parallel load shift register: truth table*

Continued on p. 152

Practical Exercise 7.7 *(Continued)*

3 Load into the register the input data $ABCD = 0111$, remembering that a logic 0 at the present input of a particular flipflop will give a logic 1 at the Q output of that flipflop. The output should be as the second line in the truth table.

Notice that the output states take place immediately the inputs are loaded, with no need for a clock pulse (and with no delay).

4 Before applying the first clock pulse, make sure that all the parallel data inputs are now at logic 1. The outputs should still be as the second line.

5 Apply four clock pulses. You should see the data shift to the right as shown in the truth table. Notice that the fourth clock pulse causes the data to leave the register.

Notice that the J and K inputs of flipflop A are unconnected (floating). If the data does not successfully shift to the right, it could be that J is taking up a logic 1 level and causing the flipflop to be set to logic 1.

The solution is to connect J to logic 0 and K to logic 1, whereby flipflop A output will reset to logic 0 after the original data bit has been shifted.

6 Clear the register, apply a combination of preset inputs of your choice, shift the data through the register, and draw up the appropriate truth table.

7 Connect the Q and $\overline{Q}$ outputs of flipflop D to inputs J and K, respectively, of flipflop A. These feedback links should now cause the data to recirculate in the register, that is, instead of leaving the right-hand side of the register for ever, the data should reappear on the left-hand side. Check that this is so by repeating procedures 1 to 3 and noting that this time, after four clock pulses, the original data (1000) is present at the register output.

The universal shift register

There are applications in which it is necessary to shift data, not only from left to right (Q_A to Q_D), but also from right to left (Q_D to Q_A). Practical Exercise 7.8 deals with the universal shift register, which can be made to shift in either direction with either serial or parallel loading: a truly all-singing, all-dancing device!

Practical Exercise 7.8

To investigate the universal shift register
For this exercise you will need the following components and equipment:

1 – 74LS194 ic (4-bit universal shift register with positive edge triggering)
4 – LED (5 mm) and resistor (270 Ω)
1 – +5 V DC power supply

The pin connections for the 74194 are shown in Figure 7.23. To determine the shift direction, the necessary mode control connections are shown in Figure 7.24.

The LED indicators and their resistors should be connected, as before, to the Q outputs.

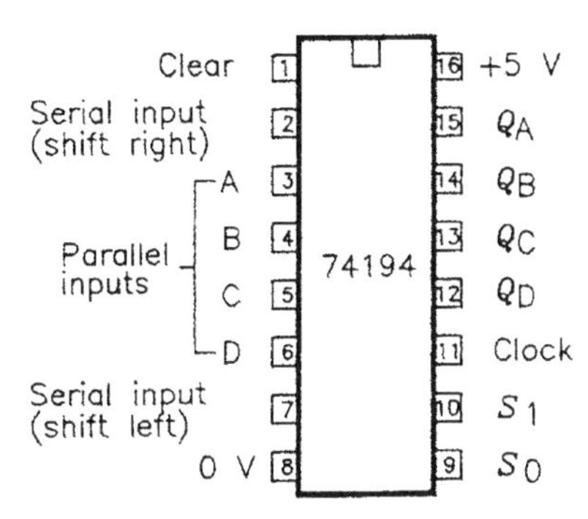

Figure 7.23 *The 74194 universal shift register: pin connections*

Action	Mode control S_0	S_1
Parallel load	1	1
Shift to the right	1	0
Shift to the left	0	1
Inhibit	0	0

Figure 7.24 *The 74194 universal shift register: mode control*

Procedure (a): serial in–parallel out

1. Clear the register by connecting the clear input to logic 0 and then, to avoid any ambiguity, to logic 1.
2. Set the parallel inputs (ABCD) to 0000.
3. To shift right:
   (a) inhibit the **shift left** serial input by connecting it to logic 0;
   (b) connect S_0 to logic 1 and S_1 to logic 0;
   (c) set the **shift right** serial input to logic 1, apply four clock pulses and notice that the data moves through the register from Q_A to Q_D.

Continued on p. 154

Practical Exercise 7.8 *(Continued)*

4 To shift left:

(a) inhibit the **shift right** serial input by connecting it to logic 0;
(b) connect S_0 to logic 0 and S_1 to logic 1;
(c) set the **shift left** serial input to logic 1, apply four clock pulses and notice that the data moves through the register from Q_D to D_A.

5 To inhibit:

(a) connect S_0 and S_1 to logic 0;
(b) set the **shift left** serial input to logic 1, apply four clock pulses and notice that there is no movement of data. Repeating this for shift right will produce the same result.

This should help to emphasize the fact that the mode connections control the operation of the register.

You should have noticed that the clear (reset) input, when enabled by logic 0, overrides all other inputs: that is, the register can be reset to 0000 at any time, irrespective of the state of the outputs.

Procedure (b): parallel in–parallel out

1 Clear the register, as before.
2 Connect S_0 and S_1 for parallel loading, set the parallel inputs (ABCD) to 1100 and apply one clock pulse. The outputs $Q_A Q_B Q_C Q_D$ should now be 1100.
3 At this point, notice that applying further clock pulses causes nothing to happen! Why is that?
(It is because the need to connect for parallel loading and the setting of the parallel outputs must come before the decision on shifting left or right): and so ...
4 Shift right (S_0 to logic 1 and S_1 to logic 0). Apply four clock pulses and notice that the data moves through the register (from *A* to *D*). Note the contents of the register after each clock pulse.
5 Repeat the setting up procedures 1 to 3.
6 Shift left (S_0 to logic 0 and S_1 to logic 0).
Apply four clock pulses and notice that the data moves through the register (from *D* to *A*). Note the contents of the register after each clock pulse.

Conclusions

1 From the results in procedure (b), copy out and complete the table in Figure 7.25. Remember: *A* is the least significant bit and, as far as the binary table is concerned, it must appear on the right-hand side.
2 What mathematical operation is performed by

(a) shifting right?
(b) shifting left?

Condition		Output: Binary D C B A	Output: Decimal
Shift right	initial	0 0 1 1	3
	first clock pulse		
	second clock pulse		
	third clock pulse*		
	fourth clock pulse*		
Shift left	initial	1 0 0 0	8
	first clock pulse		
	second clock pulse		
	third clock pulse		
	fourth clock pulse		

* Imagine the existence of outputs *E* and *F*

Figure 7.25 *The universal shift register: conclusions for Practical Exercise 7.8*

A final note on shift registers: the exercises here have concentrated on 4-bit devices. There are, however, 5, 8 and 16-bit registers available in the TTL 7400 series (and even 64-bit in CMOS 4000 series!) – all for specific purposes.

Questions

7.5 Explain how you would load (enter) the data $ABCD = 1001$ into a 4-bit shift register, which has previously been cleared to 0000.

7.6 A 4-bit shift register reads $ABCD = 1011$. List the contents of the register (as *ABCD*) shifting right, after

(a) one clock pulse
(b) two clock pulses
(c) three clock pulses
(d) four clock pulses.

7.7 Explain the process of binary multiplication when using a shift left register.

7.8 What advantage does parallel loading have over serial loading for a shift register?

Continued on p. 156

Questions (*Continued*)

7.9 A certain 4-bit shift register uses a clock frequency of 4 MHz. What time is needed to load the register (a) in series (b) in parallel? How many clock pulses are needed for each method respectively?

7.10 With reference to the universal shift register, explain the importance and use of the mode control.

7.11–7.13 Complete the waveform (timing) diagrams shown in Figures 7.26 and 7.27.

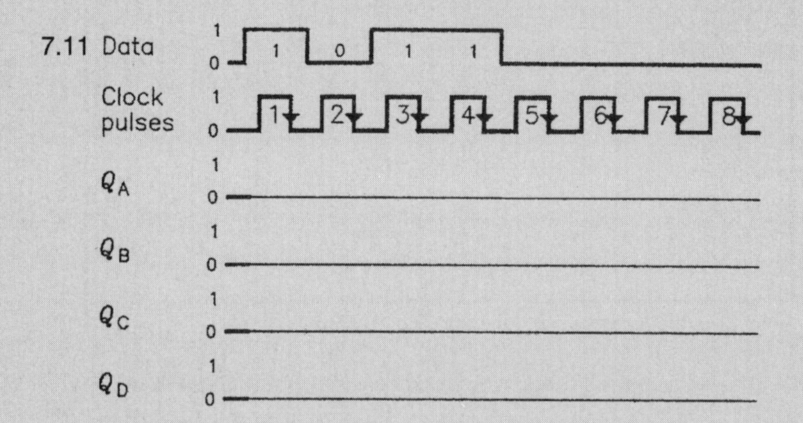

How many clock pulses does it take for:
(a) the data ABCD = 1101 to be fully stored in the register
(b) the register to be empty?

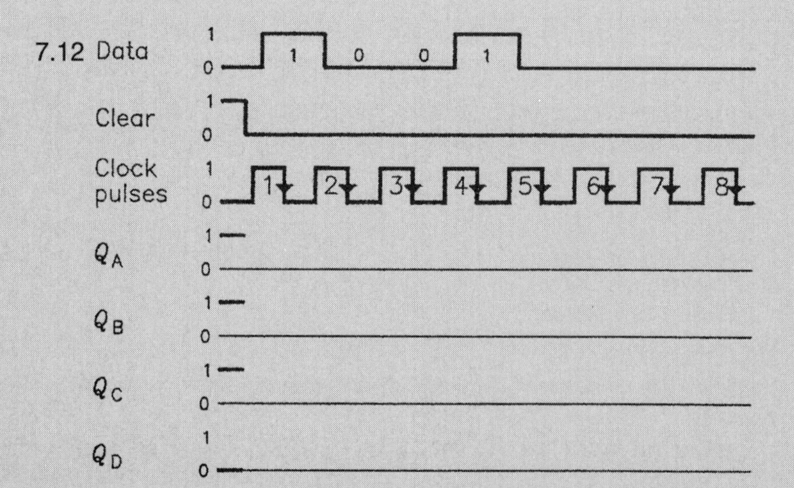

Figure 7.26 *Waveforms for Questions 7.11 and 7.12*

7.13 The following inputs are applied to the 3-bit parallel load non-recirculating shift register. Draw the waveforms of the three outputs.

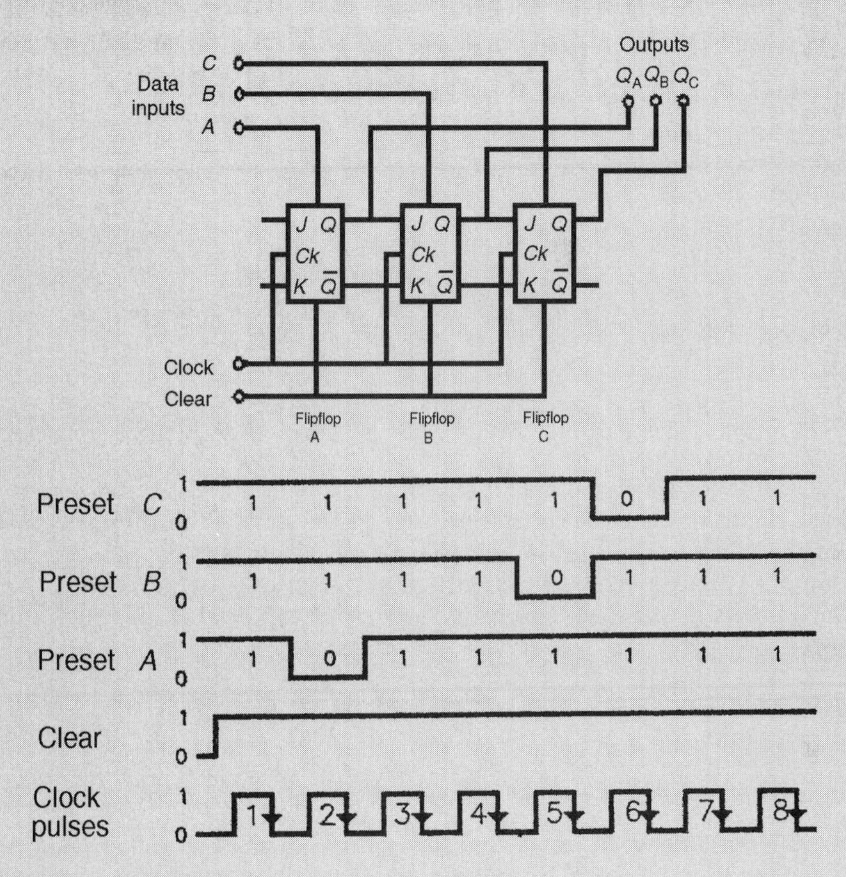

Figure 7.27 *Waveforms for Question 7.13*

8 Schmitt triggers and multivibrators

The Schmitt trigger

The Schmitt trigger is a very useful device which basically acts as a fast electronic switch. The output voltage can have one of two possible values; the action is as follows:

(a) the output voltage will switch rapidly from a low to a high level when the input voltage is greater than a certain positive-going threshold value (or **upper threshold point**);
(b) the output voltage will switch rapidly from a high to a low level when the input voltage is less than a certain negative-going threshold value (or **lower threshold point**).

Important points

- The upper and lower threshold points have different numerical voltage values.
- This difference between upper and lower threshold is a so-called **dead band**, where the input will be changing and nothing will happen at the output. This condition is known as backlash or, more technically, **hysteresis**.

Its behaviour as a fast-acting switch enables the Schmitt trigger to be used for:

(a) level detection (lighting levels being a particular example)
(b) re-shaping pulses with poor edges (long rise and fall times)
(c) providing a square-wave output signal from a sine-wave input signal
(d) eliminating noise
(e) pulse de-bouncing

The diagrams in Figure 8.1 illustrate the operation of the Schmitt trigger and the circuit symbol is shown in Figure 8.2.

Although the Schmitt trigger can be made up using discrete components (two transistors are needed), it is more conveniently available in integrated circuit form, as Practical Exercise 8.1 will demonstrate.

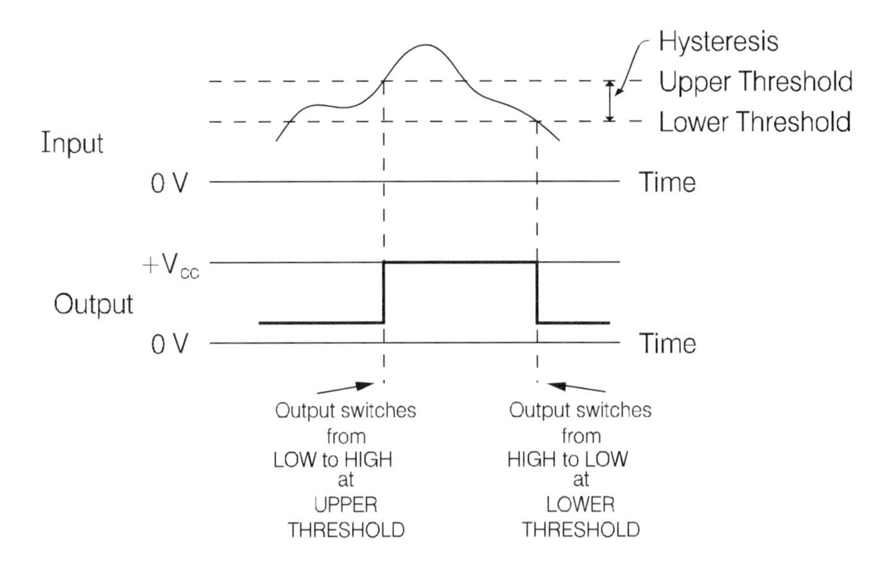

Figure 8.1 *The Schmitt trigger: waveforms to show basic operation*

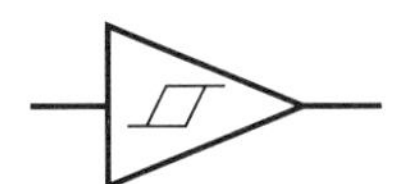

Figure 8.2 *The Schmitt trigger: circuit symbol*

Practical Exercise 8.1

To investigate the Schmitt trigger using (a) the 741 operational amplifier ic, (b) the 7414 Schmitt trigger ic
For this exercise you will need the following components and equipment:

1 – 741 ic (operational amplifier)
1 – 74LS14 ic (hex inverting Schmitt trigger) (six Schmitt triggers in one chip)
1 – ±15 V DC supply
1 – variable ±5 V DC supply
1 – resistor 2.2 kΩ
1 – resistor 10 kΩ
1 – capacitor 0.1 μF
1 – signal generator (1 kHz) with approximately 10 V peak to peak sinusoidal output voltage
1 – double-beam cathode ray oscilloscope
1 – DC voltmeter

Continued on p. 160

Practical Exercise 8.1 *(Continued)*

Procedure (a): using the 741 operational amplifier

1 Connect up the circuit of Figure 8.3. The pin connection diagram for the 741 ic is shown in Figure 8.4.

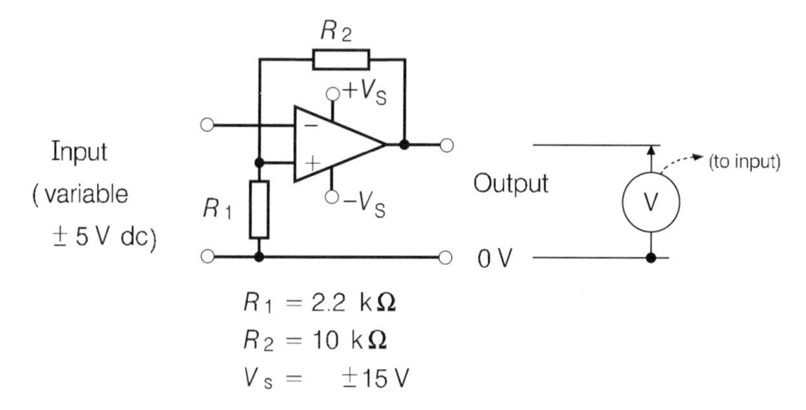

Figure 8.3 *The Schmitt trigger using the 741 operational amplifier: circuit for Practical Exercise 8.1(a)*

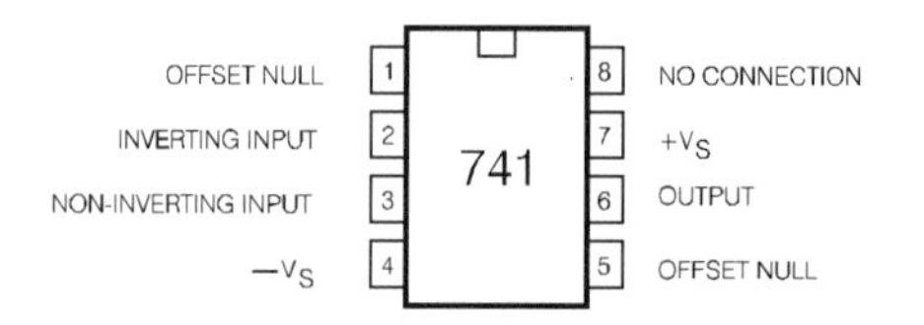

Figure 8.4 *The 741 operational amplifier: pin connections*

Notice that the 741 ic is being used as an inverting amplifier (the input is connected to the inverting '−' input of the ic but the feedback is from the output to the non-inverting '+' input). The amplifier gain is (theoretically) infinite!

2 Set the input voltage to 0 V. The output voltage should be just below the positive supply voltage (about +13 V).

3 Increase the input voltage from 0 V towards +5 V and observe the output voltmeter. Notice that there is no change at the output until a certain input voltage is reached, at which point the output will switch to a negative voltage (about −13 V). Note the value of this upper threshold input voltage.

4 Decrease the input voltage from this upper threshold value through zero towards −5 V, and note the value of the lower threshold voltage (when the output switches back to +13 V).

5 Note that it is the inverting action of the 741 amplifier which causes the output voltage to fall at the upper threshold and rise at the lower threshold. The important point to grasp here is that a definite rapid switching action occurs at the two threshold values.

6 Replace the ±5 V DC supply with the signal generator, and the voltmeter with the oscilloscope. Arrange to display both the input and output voltages.

7 Increase the input signal voltage from the generator and notice that at a certain input level a square waveform appears at the output. Draw on graph paper the input and output waveforms, one underneath the other. The waveforms should resemble those shown in Figure 8.5.

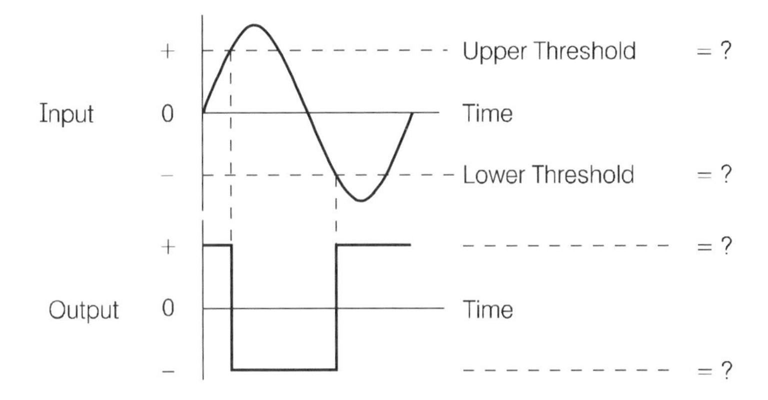

Figure 8.5 *The Schmitt trigger: waveform diagrams for the 741 operational amplifier, Practical Exercise 8.1(a)*

Make sure that the appropriate voltage levels are marked on the drawing (upper and lower threshold points for the input and the output levels).

8 It may be helpful, in completing the statements 3 and 4 which follow, to display together the waveforms at pins 2 and 3 of the 741 ic.

Conclusions

The general principles of operation of this Schmitt Trigger are now dealt with. Work through each of the statements and compare your calculated answers with those obtained from the practical measurements.

Important points

- For a supply voltage of $\pm V_S$, it is reasonable to assume that the amplifier output will be limited to a (saturated) value (V_{Osat}) of $\pm(V_S - 2)$ V. Hence if $V_S = 15$ V, $V_{Osat} = 13$ V.
- The resistors R_1 and R_2 divide the output and determine the feedback voltage applied to the non-inverting input.

Continued on p. 162

Practical Exercise 8.1 *(Continued)*

Copy out and complete the following statements:

1 Upper threshold point (UTP)

$$= \frac{R_1}{R_1 + R_2} \times (+)\ V_{Osat}\ V = +\_\_\_\_ \text{ V}$$

From the practical exercise, UTP = +____ V

2 Lower threshold point (LTP)

$$= \frac{R_1}{R_1 + R_2} \times (-)\ V_{Osat}\ \text{V} = -\_\_\_\_ \text{ V}$$

From the practical exercise, LTP = −____ V
See Figure 8.5.

Copy out the following items 3, 4 and 5, select the correct word from within the square brackets and fill in the spaces with values or names as appropriate.

3 When the inverting input becomes more [positive/negative] than the non-inverting input, the output switches rapidly to − ____ V

4 When the inverting input becomes more [positive/negative] than the non-inverting input, the output switches rapidly to + ____ V.

5 The difference between the two switching threshold points (UTP-LTP) is called ____.

Its value for this exercise = ____ V

Procedure (b): using the 7414 hex inverting Schmitt trigger

1 Connect up the circuit of Figure 8.6. The pin connection diagram of the 7414 ic is shown in Figure 8.7.

Note (a) the addition of the 'bubble' which describes the inverting Schmitt trigger and (b) the shape inside the triangle, which resembles the well-known hysteresis loop for a ferromagnetic material.

2 Set the input signal level such that there is an output square waveform being displayed. Draw on graph paper the input and output waveforms, one underneath the other.

This time, the waveforms should resemble those shown in Figure 8.8.

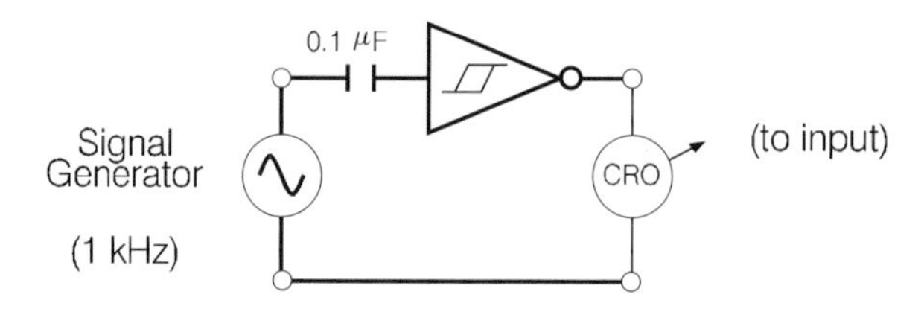

Figure 8.6 *The 7414 Schmitt trigger: circuit for Practical Exercise 8.1(b)*

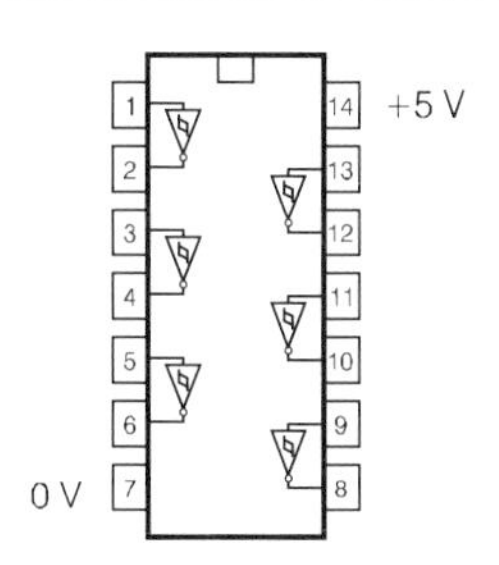

Figure 8.7 *The 7414 Schmitt trigger ic: pin connections*

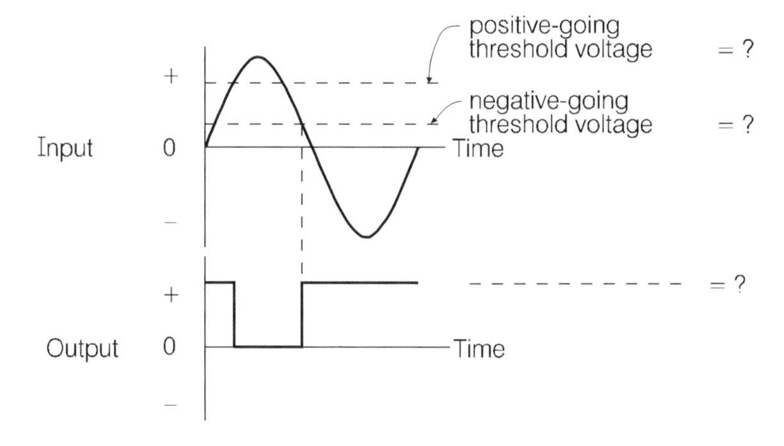

Figure 8.8 *The Schmitt trigger: waveform diagrams for Practical Exercise 8.1(b)*

Again, make sure that the appropriate voltage levels are marked on the drawing.

Conclusions

Important points

- For the ic version of the Schmitt trigger, the threshold voltages are both positive values, but are described as a **positive-going threshold voltage** and a **negative-going threshold voltage** (see Figure 8.8).

From the waveforms obtained in procedure (b) 2, estimate the following values:

(a) Positive-going and negative-going threshold voltages, comparing these values with those given in the manufacturers' data.

Continued on p. 164

Practical Exercise 8.1 *(Continued)*

(b) The voltage hysteresis.
(c) The high and low switching voltages (these were referred to in the opening paragraph of this chapter). The difference between these values will give the actual voltage height of the pulse.
(d) The duration of the output pulse.

Finally, to illustrate the hysteresis effect in diagram form, see Figure 8.9.

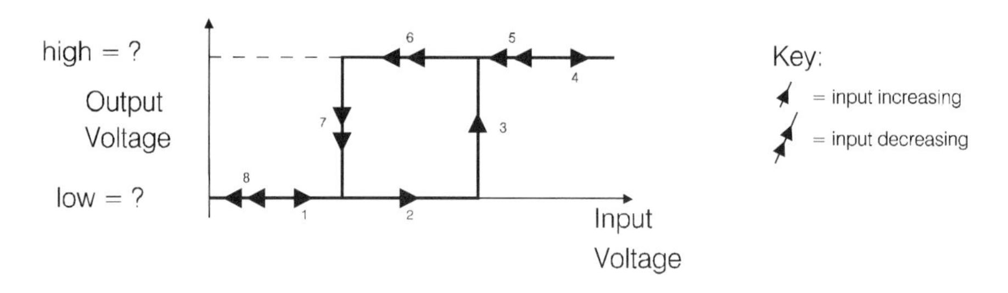

Figure 8.9 *The Schmitt trigger: hysteresis effect*

Questions

Refer to Figure 8.3.

8.1 Resistor R_1 is changed to 1 kΩ. Calculate the new values of UTP, LTP and the hysteresis.
8.2 A hysteresis value of 2 V is required. Calculate suitable values for R_1 and R_2.
8.3 For a fixed value of R_2 how does the hysteresis depend upon the value of R_1?
8.4 How does the output frequency compare with the input frequency?

Multivibrators

Multivibrators belong under the general heading of *oscillators*, whose function is to generate an output voltage of a particular shape and frequency. The multivibrator family provides rectangular waveforms in general and square waves in particular (clock pulses for computers etc.).

A rectangular waveform is one which switches between two voltage levels, which may or may not be of the same polarity. In most cases they probably will, and in some, one may be 0 V. It is almost certainly a requirement that the switching between voltage levels occurs as rapidly as possible.

The word 'multivibrator' means literally 'multiple vibrations or frequencies'. An analysis (not needed here!) would show that a square wave of (say) frequency 1 kHz consists of sine waves of frequency 1 kHz (the fundamental) together with a series of odd harmonics (three times, five times the fundamental etc.).

Main categories of multivibrator

(a) The bistable or flipflop, which has already been met, has two stable states and will remain in one or the other until it is triggered into the opposite state.

(b) The astable or free-running vibrator provides a continuous train of pulses (either rectangular or square).

Astable means *not stable*: the device switches from one state to the other, staying in each state for a length of time determined by CR circuit component values. It has no stable state. It finds application as a clock pulse generator.

(c) The monostable or one shot has one stable state (and thus one unstable state). It rests in this stable state until triggered into the unstable condition, in which it will remain for a period of time determined, again, by the component values. Its use is as a generator of pulses of controlled width and fixed amplitude.

The 74121 monostable is a non-retriggerable multivibrator, which means that no further trigger pulse after the first will be effective until the monostable has been reset. The retriggerable type (74123 is an example), which responds to all trigger pulses whenever they occur, enables the pulse width to be extended as required.

The action of these three types is illustrated by the waveforms in Figure 8.10.

The astable and monostable multivibrators will now be dealt with by way of practical exercises.

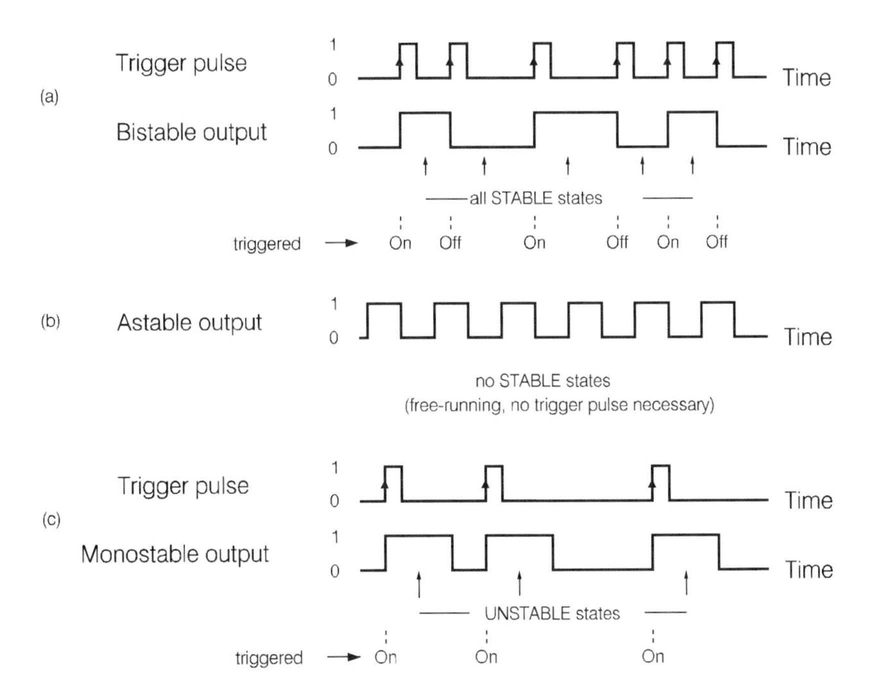

Figure 8.10 *Waveforms to illustrate the three types of multivibrator*

Practical Exercise 8.2

To investigate the astable multivibrator using (a) the 741 operational amplifier ic, (b) the 555 timer ic, (c) the 7414 Schmitt trigger ic
For this exercise you will need the following components and equipment:

1 – 741 ic (operational amplifier)
1 – 555 ic (timer)
1 – 74LS14 ic (hex inverting Schmitt trigger)
1 – ±15 V DC supply
1 – ±5 V DC supply
1 – set of components for the 741 circuit:
  resistors 1 × 100 kΩ, 1 × 10 kΩ, 1 × 1 kΩ
  potentiometer 1 × 5 kΩ
  capacitor 1 × 0.1 μF
1 – set of components for the 555 circuit:
  resistors 2 × 4.7 kΩ
  capacitors 1 × 0.01 μF
1 – set of components for the 7414 circuit:
  resistor 1 × 1 kΩ
  capacitor 1 × 0.1 μF
1 – double-beam cathode ray oscilloscope

Procedure (a): using the 741 operational amplifier

1 Connect up the circuit of Figure 8.11. The pin connection diagram for the 741 ic is shown in Figure 8.4.

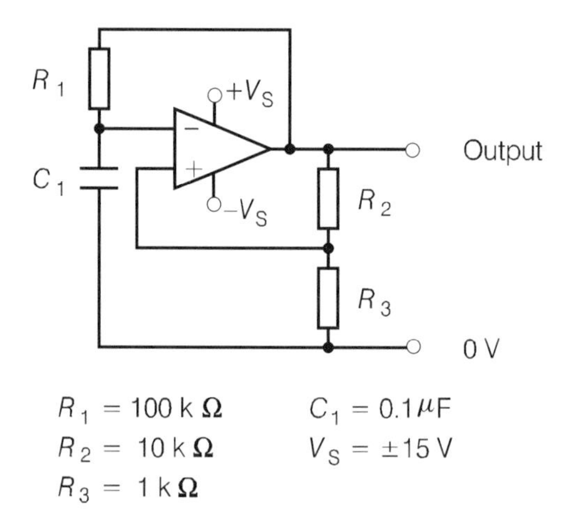

Figure 8.11 *The astable multivibrator using the 741 operational amplifier: circuit for Practical Exercise 8.2(a)*

2 Sketch, on graph paper, the waveform of the voltages across the capacitor C_1 and at the output, which should resemble those shown in Figure 8.12. Make sure that you show clearly the voltage and time values.

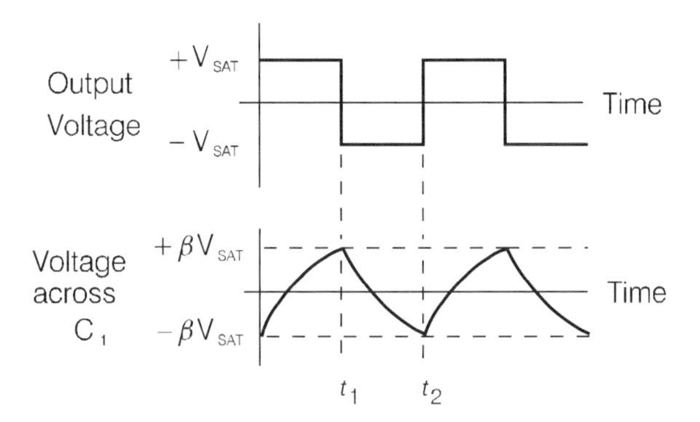

Figure 8.12 *The 741 astable multivibrator: typical waveforms for Practical Exercise 8.2(a)*

3 Measure (using the oscilloscope) the periodic time of the output waveform and hence calculate its frequency.
4 Modify the circuit to that of Figure 8.13 and show that a variation in frequency can be achieved. Measure this variation (maximum to minimum).

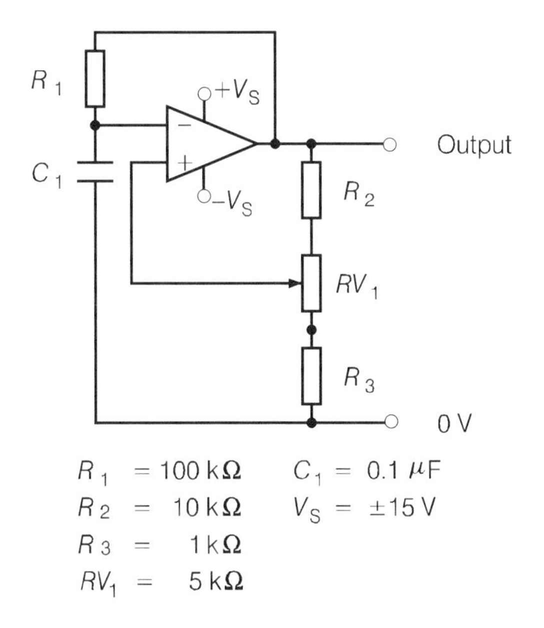

Figure 8.13 *The 741 astable multivibrator, variation of frequency: circuit for Practical Exercise 8.2(a)*

Continued on p. 168

Practical Exercise 8.2 *(Continued)*

Conclusions

If you now work through the following exercise, it should help your understanding of the action of this form of astable.

Important points

- From the circuit (Figure 8.11), notice the positive feedback from the output, through the potential divider R_2 and R_3 to the non-inverting (+) input.
- For a supply voltage V_s of ± 15 V, the maximum (saturated) output from the amplifier (V_{Osat}) will be about ± 13 V.
- The fraction of the output voltage fed back (β) is given by

$$\beta = \frac{R_3}{R_2 + R_3} = \frac{1\ \text{k}\Omega}{11\ \text{k}\Omega} = 0.091\ \text{k}\Omega$$

- The expected voltage at the non-inverting input for the saturated output will be given therefore by

$$\beta \times \pm V_s$$

which, for the above values, gives

$$0.091 \times \pm 13\ \text{V} \quad \text{or} \quad \pm 1.18\ \text{V}$$

Copy out the following statements, selecting the correct word from within the square brackets and filling in the spaces with values or names as appropriate.

1 Assume that the output is at its maximum positive value, that is, the output voltage $= +V_{Osat} = +$____.
This positive voltage charges up the capacitor (C_1) through the ____ ( ).

2 As C_1 charges up positively, the voltage at the ____ input [rises/falls] until after a time $t = t_1$ when it becomes more [positive/negative] than the voltage at the ____ input.
The amplifier output will now swing rapidly [negative/positive] to a value of ____ V, helped by the action of [positive/negative] feedback.

3 This [negative/positive] voltage will again charge up the ____ ( ) through the ____ ( ), until the ____ input becomes more [positive/negative] than the ____input, at which point ($t = t_2$), the amplifier output voltage switches back to a [positive/negative] saturated value (= ____ V).

4 The theoretical value of the frequency (f) of the output waveform, also known as **pulse repetition frequency** (prf), is given by $f = 1/T$, where

$$T = 2C_1R_1 \log_e \left(1 + \frac{2R_3}{R_2}\right)$$

(f in hertz, T in seconds, C in farads, R in ohms, and e the base of Naperian logarithms).

Compare the theoretical and measured values and account for any differences.

5 For the circuit modification shown in Figure 8.13 calculate the theoretical frequency range possible.

Note. You will need to take into account the value of RV_1 that is in series with R_2 and R_3 respectively.

Procedure (b): using the 555 timer

1 Connect up the circuit of Figure 8.14. The pin connection diagram for the 555 ic is shown in Figure 8.15.

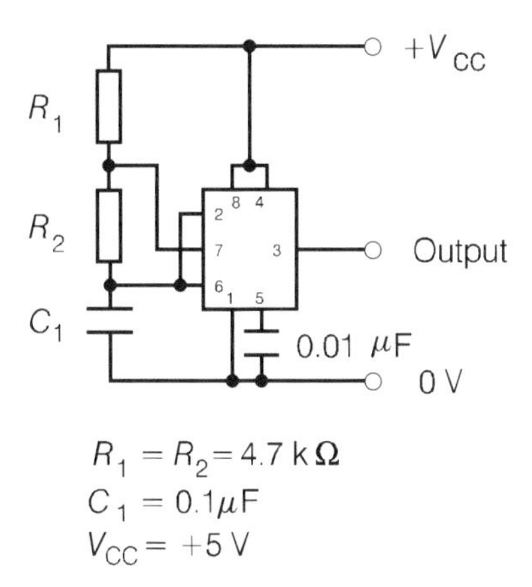

Figure 8.14 *The astable multivibrator using the 555 timer: circuit for Practical Exercise 8.2(b)*

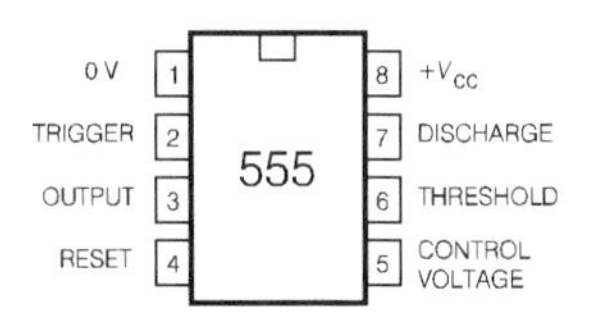

Figure 8.15 *The 555 timer: pin connections*

2 Sketch, on graph paper, the waveform of the voltages across the capacitor C_1 and at the output, which should resemble those shown in Figure 8.16. Make sure that you show clearly the voltage and time values.

3 Measure (using the oscilloscope) the periodic time of the output waveform and hence calculate its frequency.

Continued on p. 170

Practical Exercise 8.2 (*Continued*)

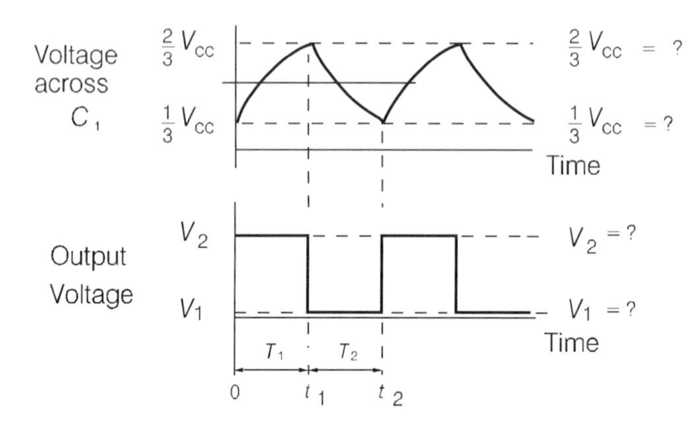

Figure 8.16 *The 555 astable multivibrator: typical waveforms for Practical Exercise 8.2(b)*

Conclusions

Important points

- At switch-on, the capacitor is uncharged and the output (at pin 3) will be high (approximately $+V_{CC}$).
- Capacitor C_1 charges up towards the supply voltage $+V_{CC}$ with a time constant of $C \times R$ seconds, where

  $$C = C_1 \text{ (farads)} \quad \text{and} \quad R = (R_1 + R_2) \text{ (ohms)}$$

- When the voltage across C_1 (pin 6) reaches an upper threshold value of $\frac{2}{3}V_{CC}$ (at time t_1 in Figure 8.16) the output goes low (approximately 0 V).
- Capacitor C_1 then discharges through R_2 (time constant C_1R_2) until its voltage reaches the lower threshold of $\frac{1}{3}V_{CC}$ (at time t_2). At this point, the output switches back to its high level.
- Using a supply voltage (V_{CC}) of +5 V provides a TTL-compatible output (pin 3).

Calculate the following values.

(a) Time T_1 for which the output is high (called the **Mark**) is given by

$$T_1 = 0.7(R_1 + R_2)C_1$$

(R_1 and R_2 in ohms, C_1 in farads, T_1 in seconds).
Calculate T_1.

(b) Time T_2 for which the output is low (called the **Space**) is given by

$$T_2 = 0.7 R_2 C_1$$

Calculate T_2.

(c) Periodic time, $T = T_1 + T_2 = 0.7(R_1 + 2R_2)C_1$, and

frequency, $(f) = 1/T$

(T in seconds, f in hertz).
Calculate f.

(d) The mark–space ratio is found from

$$\frac{\text{time for the mark}}{\text{time for the space}} = \frac{T_1}{T_2} = \frac{0.7(R_1 + R_2)C_1}{0.7 R_2 C_1} = \frac{R_1 + R_2}{R_2}$$

Calculate the mark–space ratio.

(e) The duty cycle is found from

$$\frac{\text{Time for the mark (the 'on' time)}}{\text{Time for mark and space (the total time)}}$$

$$= \frac{0.7(R_1 + R_2)C_1}{0.7(R_1 + 2R_2)C_1} = \frac{R_1 + R_2}{R_1 + 2R_2}$$

Calculate the duty cycle.

Procedure (c): using the 7414 Schmitt trigger

1 Connect up the circuit in Figure 8.17. The pin connection diagram for the 7414 ic is shown in Figure 8.7.
2 Measure the frequency of the output waveform.

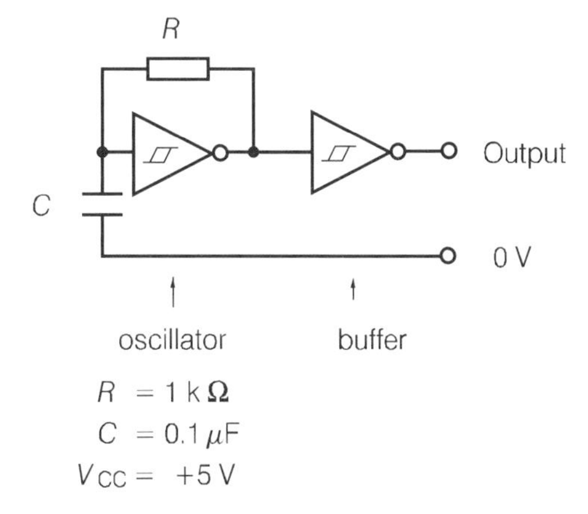

Figure 8.17 *The astable multivibrator using the 7414 Schmitt trigger: circuit for Practical Exercise 8.2(c)*

Questions

Refer to Figure 8.14.

8.5 What would be the effect on the mark–space ratio of making R_2 much larger than R_1 (say, $R_1 = 1$ kΩ and $R_2 = 10$ kΩ)?
8.6 Why can the mark–space ratio never be less than 1:1?
8.7 What would be the effect (if any) on the frequency, of having a supply voltage greater than 5 V?
8.8 By what practical method could a variable mark–space ratio output be provided?

Practical Exercise 8.3

To investigate the monostable multivibrator using (a) the 555 timer ic, (b) the 74121 monostable ic
For this exercise you will need the following components and equipment:

1 – 555 ic (timer)
1 – 74121 ic (monostable multivibrator)
1 – +5 V DC supply
1 – set of components for the 555 timer circuit:
resistors 1 × 1 kΩ, 1 × 120 kΩ, 1 × 470 kΩ, 1 × 1 MΩ
capacitors 1 × 10 nF, 2 × 0.01 μF, 1 × 10 μF
1 – set of components for the 74121 circuit:
resistor 10 kΩ
variable resistor 50 kΩ
capacitor 0.01 μF
1 – square wave generator (+5 V at 1 kHz)
1 – double beam cathode ray oscilloscope

Procedure (a)

1 Connect up the circuit of Figure 8.18(a) making $R_1 = 1$ MΩ. The pin connections for the 555 timer are shown in Figure 8.15.
Leave switch S open (this switch can be a lead which is yet to be connected). At this point the output voltage at pin 3 should be 0 V.
2 Apply a trigger pulse by momentarily closing switch S. This will then start the output pulse, with the output voltage rising to +5 V (= $+V_{CC}$).
After a time (T) given by $T = 1.1C_1R_1$, where C_1R_1 is the time constant of the circuit, the pulse will end with the output voltage returning to 0 V.
For the values chosen, with C_1 in farads and R_1 in ohms, T should be 11 seconds.

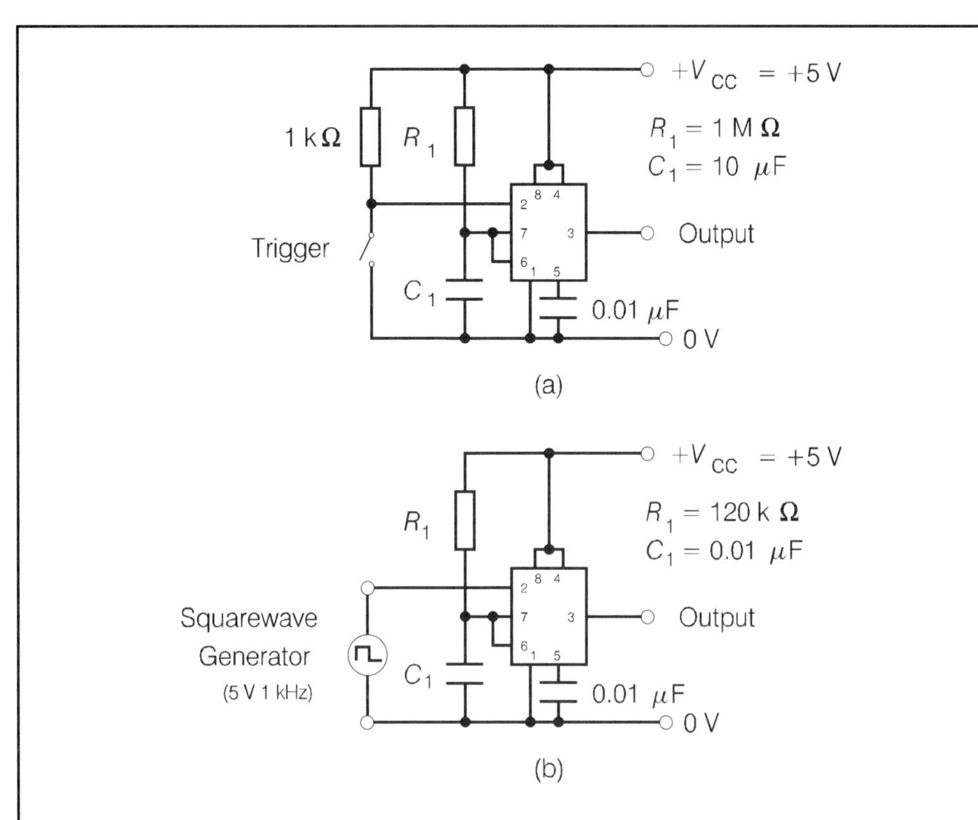

Figure 8.18 *The monostable multivibrator using the 555 timer: circuit for Practical Exercise 8.3(a)*

3 Use your watch (or count in seconds), to check the time for the output pulse. This time is also known as the **pulse length**.

4 Change R_1 to 470 kΩ and confirm that the new pulse length is approximately one-half of that previously found.
The oscilloscope can be used, if preferred, to check the output level changes. If you use it to inspect the charging waveform across the capacitor, be prepared for a problem! The oscilloscope may have an input impedance as low as 1 MΩ, and could slow down the charging and possibly prevent the monostable switching back to 0 V.

5 Take a look at questions 2 and 3 under 'Conclusions' and perform the appropriate investigation.

6 Change the circuit to that shown in Figure 8.18(b). The signal generator should be set to give a 5 V amplitude square wave at a frequency of 1 kHz. This will now provide a series of pulses from the multivibrator for visual inspection purposes.

7 Sketch, on graph paper, the waveforms of the voltage across the capacitor C_1 and at the output, which should resemble those shown in Figure 8.19. Make sure that you show clearly the voltage and time values.

Continued on p. 174

Practical Exercise 8.3 (*Continued*)

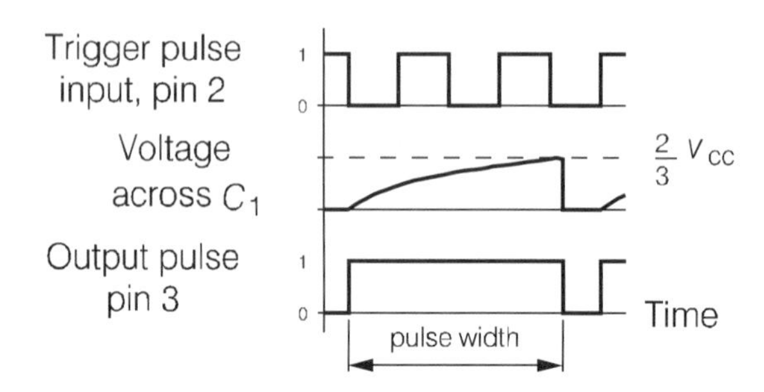

Figure 8.19 *The 555 monostable multivibrator: typical waveforms for Practical Exercise 8.3(a)*

Conclusions

Copy out the following statements, selecting the correct word from within the square brackets and filling in the spaces with values or names as appropriate. Then consider questions 1–3.

(a) Initially, with the power supply on ($V_{CC} = +5$ V) and no trigger pulse, the monostable multivibrator will rest in its [stable/unstable] state. The trigger input (pin 2) will be held [high/low] and the output (pin 3) will be [low/high].

(b) Connecting pin 2 to the 0 V line, by closing switch S, is equivalent to applying a [positive/negative] -going ____ pulse. When the trigger input falls to $\frac{1}{3}V_{CC}$, the output goes [low/high] and the monostable is set to its [stable/unstable] state.

(c) Capacitor C_1 charges towards $+V_{CC}$ until the voltage across it reaches the threshold value of $\frac{2}{3}V_{CC}$. At this point the monostable is reset to its [stable/unstable] state and the output returns to [low/high].

1 Comparing the actual pulse lengths with their theoretical values, what is likely to be the main reason for any discrepancy?
(*Hint.* What **type of capacitor** did you use?)

2 What would be the result of a trigger pulse which lasted longer than the expected time for the output pulse?

3 Once the output has been triggered, what effect would further trigger pulses have on the pulse length?

Procedure (b): using the 74121 monostable multivibrator ic

The connection diagram for the multivibrator is shown in Figure 8.20 and the pin connections for the 74121 ic are shown in Figure 8.21.

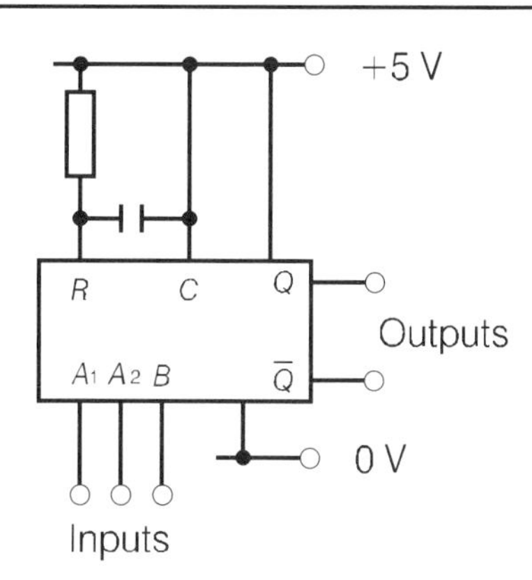

Figure 8.20 *The 74121 monostable multivibrator: connection diagram for Practical Exercise 8.3(b)*

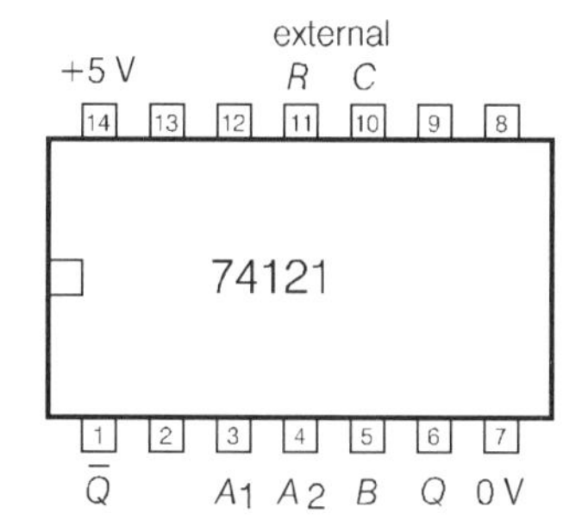

Figure 8.21 *The 74121 monostable multivibrator: pin connections*

This ic has three control inputs, A_1, A_2 and B. It requires two external timing components R and C, with the pulse width (t) being given by

$$t = 0.7RC$$

(R in ohms, C in farads, t in seconds).

The three modes of operation can now be investigated.

1. Refer to the table in Figure 8.22 and investigate each mode in turn. Use the cathode ray oscilloscope to display the input and output waveforms and measure the pulse width of the output. Draw each set of waveforms.
2. Compare theoretical and practical pulse width values.
3. Change the value of R to the 50 kΩ variable and investigate the range of pulse widths available.

Continued on p. 176

Practical Exercise 8.3 *(Continued)*

Trigger pulse to input	Inputs A_1	A_2	B	Output Q
B	0	0	↑	↑
A_1	↓	1	1	↑
A_2	1	↓	1	↑

Figure 8.22 *The 74121 monostable multivibrator: modes of operation*

Question

8.9 How would the pulse width of the monostable depend upon the supply voltage to the 555 timer?

The 555 timer ic

As a conclusion to this chapter, a summary is provided of the function of each of the terminals of the 555 timer ic, from the instant of switch-on. Figure 8.23 relates the terminals to the ic pin numbers.

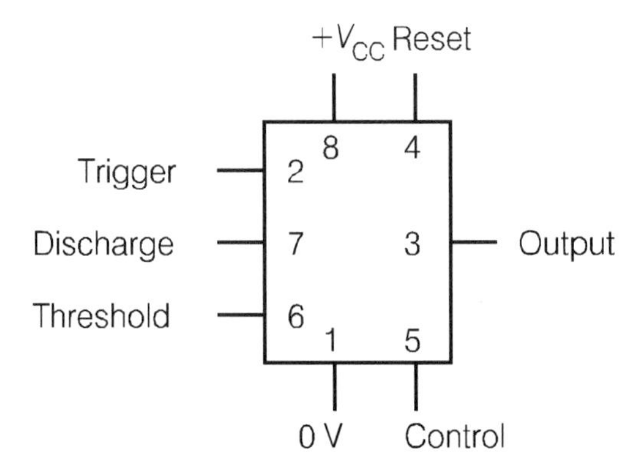

Figure 8.23 *The 555 timer ic: terminal identification*

Pin number	Title	Function
8	$+V_{CC}$	Power supply (+)
1	0 V	Power supply (−)
3	Output	In the quiescent (nothing happening) state, the voltage at this pin will be **low**, at approximately 0.4 V. The **high** level output voltage is approximately $(+V_{CC} - 1)$ V, which can source a maximum load of 200 mA.
2	Trigger	In the quiescent state, the trigger voltage will be held at **greater** than $+\frac{1}{3}V_{CC}$. When the trigger voltage is taken **below** $+\frac{1}{3}V_{CC}$, the output voltage (pin 3) switches from **low** to **high**.
6	Threshold	In the quiescent state, the threshold voltage will be held at **less** than $+\frac{2}{3}V_{CC}$. When the threshold voltage is taken **above** $+\frac{2}{3}V_{CC}$, the output voltage switches from **high** to **low**.
7	Discharge	With the discharge voltage at **0 V**, the output voltage will be **low.**
4	Reset	With a reset voltage of between 0 V and 0.4 V, the output will be forced **low**, thus overriding any previous output state. When not being used, it should be connected to $+V_{CC}$.
5	Control	Can be used to vary the timing, independently of the RC network. For most applications, it is connected by means of a capacitor to 0 V, in order to maintain good noise immunity.

Logic gate oscillators

An oscillator is a device that will produce an output voltage waveform with no apparent input beyond that of the supply voltage. The multivibrators met earlier are examples of such a device.

The requirement of oscillators is for a stable and accurate output frequency. Unfortunately, where discrete components are used, these requirements will depend on the tolerances and stability of the components, together with stability of supply voltage and ambient temperature.

A quartz crystal is a piezoelectric material and as such, can change mechanical energy into electrical energy (and vice versa). It will vibrate at its natural resonant frequency when given the appropriate energy, and in turn will provide a **sinusoidal** voltage at its output terminals.

Thus, where accuracy and stability of frequency are paramount, the quartz crystal is used. Typical applications are for computer clocks, digital watches, and many kinds of electronic equipment. A particular type of oscillator is investigated in the following Practical Exercise.

Practical Exercise 8.4

To investigate the crystal-controlled oscillator
For this exercise you will need the following components and equipment:

1 – 74LS04 ic (hex inverter) (NOT gate)
1 – general purpose crystal (5 MHz max, see note below)
2 – 1 kΩ resistor
1 – 1 kΩ dual potentiometer
1 – 10 nF capacitor
1 – +15 V variable DC power supply
1 – single beam cathode ray oscilloscope

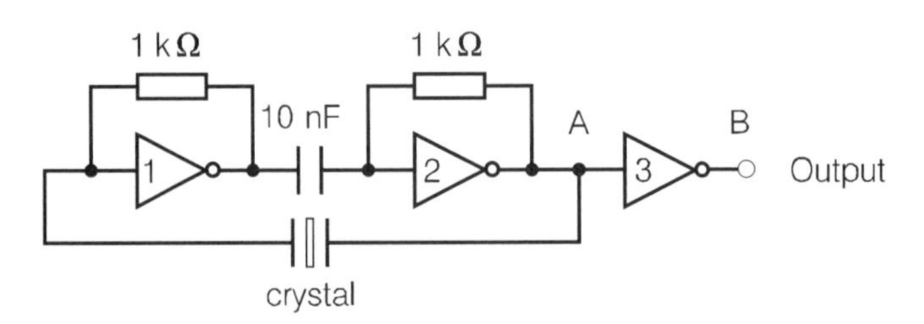

Figure 8.24 *The crystal controlled oscillator: circuit for Practical Exercise 8.4*

Note. The larger the crystal resonant frequency, the larger the CRO **bandwidth** will need to be for satisfactory frequency measurement.

Procedure

1. Connect up the circuit shown in Figure 8.24. The pin connection diagram for the 7404 ic is given in Figure 2.13. The supply voltage should be set to +5 V.
2. Measure the frequency of the output signal, which in theory **should** be sinusoidal, but in practice may be slightly distorted.
3. Measure the signal peak-to-peak output voltage at points A and B.
4. Replace the two 1 kΩ fixed resistors with the dual potentiometer wired up as two variable resistors. Adjust their value and observe if any improvement (less distortion) in the shape of the output waveform can be achieved.
5. With the supply voltage set to +15 V, repeat the measurement of frequency.

Conclusions

1 If you found a difference in the voltages at points A and B, give a likely reason for this. Hence explain the purpose of incorporating gate 3.
2 Comment on the change of supply voltage in its effect on the frequency of the output voltage.
3 The usual requirement of a crystal oscillator in digital systems is to provide a rectangular voltage waveform. Explain briefly how the conversion from sinusoidal to rectangular may be achieved. *Hint.* Refer to earlier work!
4 Further to conclusion 3, why is it not essential for the oscillator output to be perfectly sinusoidal to start with?

9 MSI combinational logic circuits

This chapter returns to the topic of combinational logic circuits, in which, by way of a reminder, the state of the output(s) is determined by the state of the present inputs.

The logic units discussed here, namely multiplexers, demultiplexers, code converters and binary adders, can, very largely, be made up using individual gates. In practice this effort is unnecessary since these units are all available as complete integrated circuits, in MSI (medium scale integration) form.

Multiplexers

A multiplexer (often abbreviated to MUX) is the electronic equivalent of a very fast-acting, single-pole, multi-way rotary switch. The principle is that a number of input channels are connected, one at a time, to a single output line, with each selected input channel being connected for a specified period of time. The process is known as time-division multiplexing (TDM). See Figures 9.1 and 9.2.

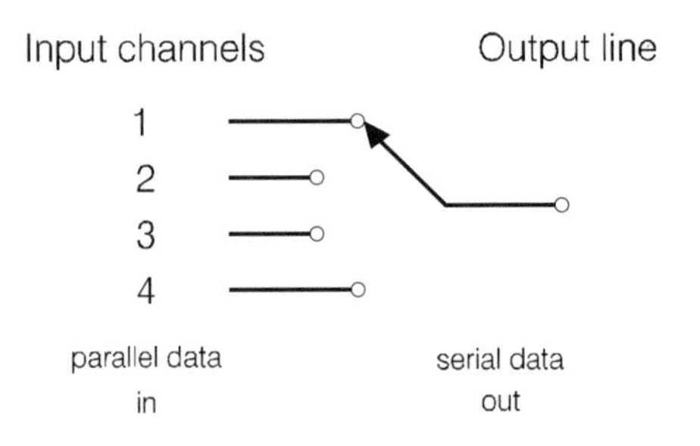

Figure 9.1 *The multiplexer as a rotary switch*

The input channels are selected by a channel addressing system, using a binary coded signal. For example, four input channels would require a 2-bit address (giving $2^2 = 4$ lines), as shown in Figure 9.3; eight inputs would need a 3-bit (for $2^3 = 8$ lines) address and so on.

Since the inputs are in parallel and the single output line corresponds to a series (or serial) line, one application of the multiplexer is as a **parallel-to-serial converter**.

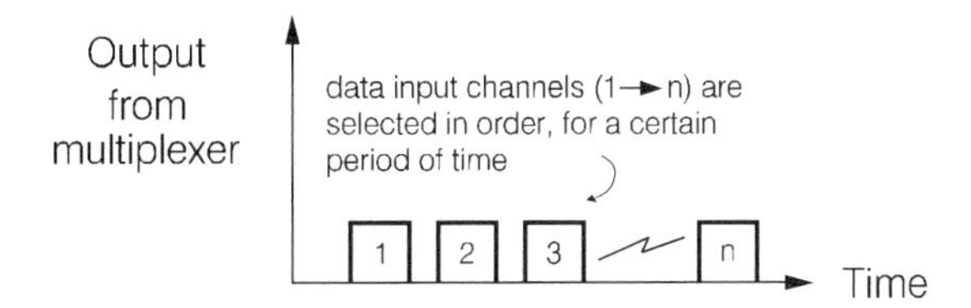

Figure 9.2 *Time-division multiplexing*

Input channel	Channel address B	A
1 (D_0)	0	0
2 (D_1)	0	1
3 (D_2)	1	0
4 (D_3)	1	1

Figure 9.3 *Channel addressing*

This action of selecting one data channel from a number of available channels gives the multiplexer an alternative name of **data selector**. As an example, a four-channel multiplexer can be also described as either a 4-line to 1-line data selector or a 1-of-4 data selector.

The action of the 4-input multiplexer is illustrated by the waveforms in Figure 9.4.

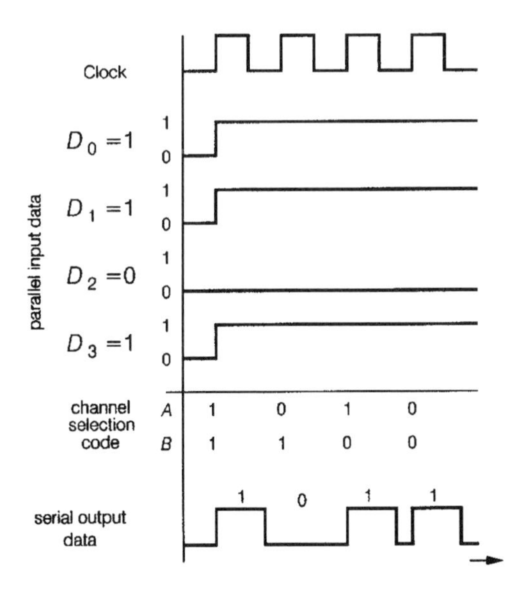

Figure 9.4 *Waveforms to show the action of the 4-input multiplexer*

Practical Exercise 9.1

To investigate the multiplexer
For this exercise you will need the following components and equipment:

1 – 74LS153 ic (dual 4-to-1 multiplexer)
1 – +5 V DC power supply
1 – LED (5 mm) and series resistor (270 Ω)

The pin connection diagram for the 74153 ic is shown in Figure 9.5. The following details may be helpful:

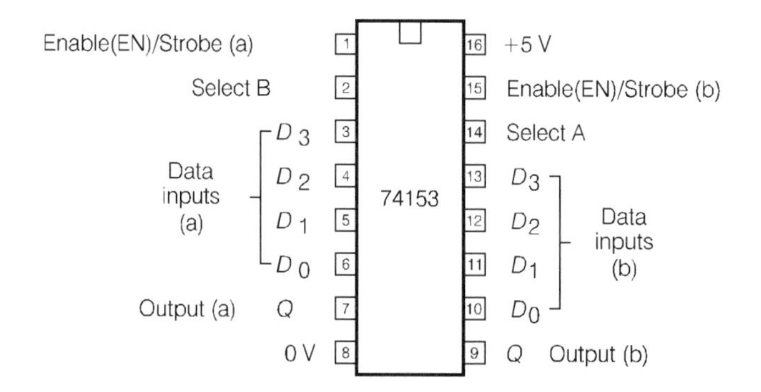

Figure 9.5 *The 74153 dual 4-line to 1-line multiplexer: pin connections*

The two separate multiplexers are referred to as (a) and (b).

D_0 to D_3 are the four data inputs. *Q* is the output. *A* and *B* are the two channel selection inputs.

With the *EN* (**enable** or **strobe**) input set at logic 1, the *Q* output will be logic 0. With the *EN* input at logic 0 (i.e. enabled), the *Q* output will take the logic level of the selected input.

Procedure

1 Connect up the circuit of Figure 9.6, which will use multiplexer(a).
2 Set the data word to $D_0D_1D_2D_3 = 1101$.
3 Work through the truth table (a), Figure 9.7, completing the column for the output *Q*, with the enable input (*EN*) initially at logic 1 and then at logic 0.
4 Work through the truth table (b), Figure 9.8, with the data inputs as shown.

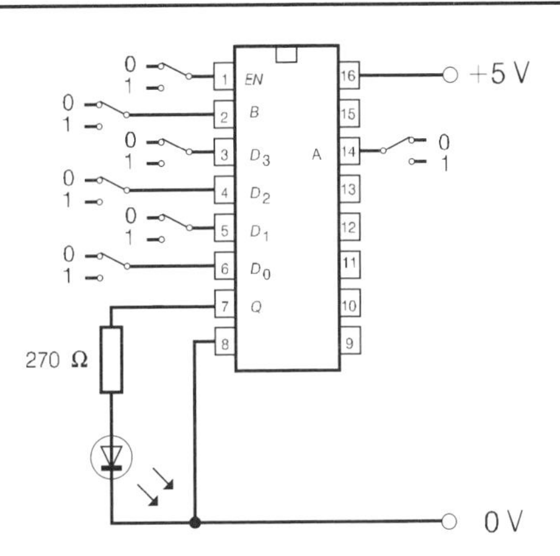

Figure 9.6 *The 74153 multiplexer: circuit for Practical Exercise 9.1*

data	channel select		EN at logic 1 output	EN at logic 0 output	output
	B	A	Q	Q	Z
$D_0 = 1$	0	0			$\bar{A}.\bar{B}.D_0$
$D_1 = 1$	0	1			$A.\bar{B}.D_1$
$D_2 = 0$	1	0			——
$D_3 = 1$	1	1			$A.B.D_3$

Figure 9.7 *Truth table (a) for Practical Exercise 9.1*

data	channel select		data inputs				(EN at logic 0) output
	B	A	D_3	D_2	D_1	D_0	Q
D_0	0	0	X	X	X	1	
D_0	0	0	X	X	X	0	
D_1	0	1	X	X	1	X	
D_1	0	1	X	X	0	X	
D_2	1	0	X	1	X	X	
D_2	1	0	X	0	X	X	
D_3	1	1	1	X	X	X	
D_3	1	1	0	X	X	X	

Where an X is shown, it means that the logic level does NOT matter

Figure 9.8 *Truth table (b) for Practical Exercise 9.1*

Important points

- Each of the four lines of the truth table in Figure 9.7 represents the address of one of the data inputs.
- While that particular data input is being addressed, the logic level of the other data inputs is unimportant, that is, they are in a don't care/doesn't matter state, usually shown by an 'X' in the truth table (Figure 9.8).
- The output conditions of the truth table can be written as a Boolean expression. Hence from Figure 9.8 (and also from Figure 9.7) we can write

  $$Z = \overline{A} \cdot \overline{B} \cdot D_0 + A \cdot \overline{B} \cdot D_1 + \overline{A} \cdot B \cdot D_2 + A \cdot B \cdot D_3$$

- Multiplexers can be used to implement a Boolean expression. For example, suppose that we wish to implement the logic statement given by

  $$Z = \overline{A} \cdot \overline{B} + A \cdot B$$

data line	channel select B	A	output Q	Z
D_0	0	0	1	$\overline{A}.\overline{B}.D_0$
D_1	0	1	0	
D_2	1	0	0	
D_3	1	1	1	$A.B.D_3$

Figure 9.9 *Multiplexer truth table for* $Z = \overline{A} \cdot \overline{B} + A \cdot B$

The truth table is given in Figure 9.9 and shows the two input conditions that will produce a logic 1 output, namely when data inputs D_0 and D_3 equal 1. Note that the complete expression for Z is given by

$$Z = \overline{A} \cdot \overline{B} \cdot D_0 + A \cdot B \cdot D_3$$

The implementation is shown in Figure 9.10 and can be tested in the next practical exercise.

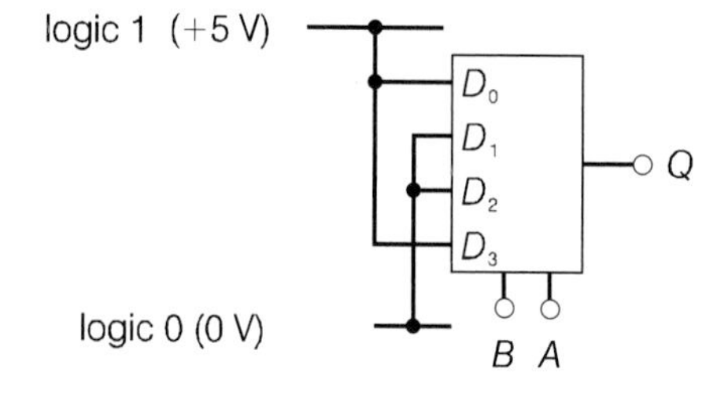

Figure 9.10 *Using the multiplexer to implement a Boolean function*

Practical Exercise 9.2

To use the multiplexer to implement a Boolean function
For this exercise you will need the following components and equipment:

1 – 74LS153 ic (dual 4-to-1 multiplexer)
1 – +5 V DC power supply
1 – LED (5 mm) and series resistor (270 Ω)

Procedure

1 Connect up the circuit of Figure 9.6, referring if necessary to Practical Exercise 9.1 for the setting-up information.
2 To implement the expression $Z = \overline{A} \cdot \overline{B} + A \cdot B$, set the data inputs to $D_0 D_1 D_2 D_3 = 1001$. Confirm that the output Q agrees with the truth table in Figure 9.9.
3 Draw up the multiplexer truth table for the expression

$$Z = A \cdot \overline{B} + \overline{A} \cdot B$$

and give the logic level for each data input.
4 Set up the necessary conditions and check the truth table by practical investigation.

Conclusions

1 For the Boolean expression $Z = A \cdot B \cdot \overline{C} + \overline{A} \cdot \overline{B} \cdot C$:

(a) Draw up the truth table.
(b) Draw the logic diagram using AND, OR and NOT gates.
(c) Estimate the total number of ic's required for the implementation of (b). (Assume the following ic availability: quad 2-input AND, triple 3-input AND, quad 2-input OR, hex inverter – all in quantities as necessary.)
(d) State the logic level of each data input for implementation using a 1-of-8 data selector.

2 Repeat the procedure of conclusion 1 for the expression

$$Z = A \cdot B \cdot C \cdot D + \overline{A} \cdot \overline{B} \cdot \overline{C} \cdot \overline{D} + A \cdot \overline{B} \cdot \overline{C} \cdot D + A \cdot B \cdot \overline{C} \cdot \overline{D}$$
$$+ \overline{A} \cdot B \cdot C \cdot \overline{D} + \overline{A} \cdot B \cdot \overline{C} \cdot D + \overline{A} \cdot \overline{B} \cdot C \cdot D$$

The requirement this time is for a 1-of-16 multiplexer.
3 As a result, state what you consider to be the advantage(s) in using a multiplexer as opposed to a number of logic gates.

Question

9.1 Use the multiplexer technique to implement the functions

(a) $Z = \overline{A} \cdot \overline{B} + B \cdot C$

(b) $Z = A \cdot B + \overline{A} \cdot \overline{C}$

Demultiplexers

A demultiplexer (DMUX) performs the opposite function of a multiplexer, that is, it returns the data to its original form. Thus the input data from the single channel is connected for a specified period of time to a selected output line using the address code previously mentioned. See Figure 9.11.

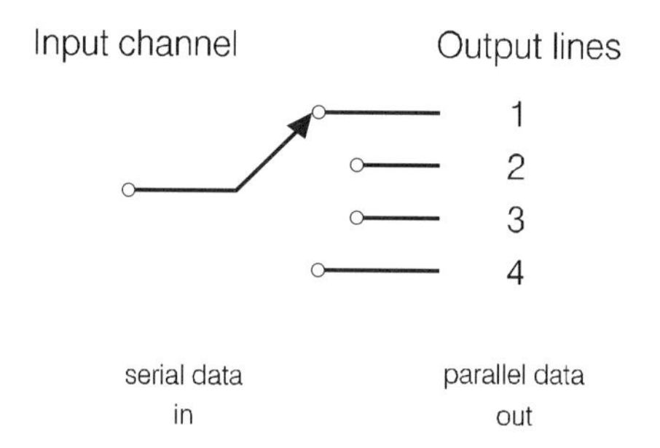

Figure 9.11 *The demultiplexer as a rotary switch*

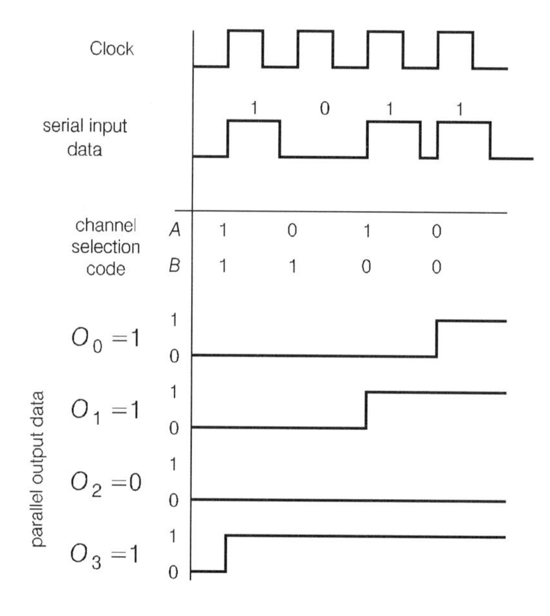

Figure 9.12 *Waveforms to show the action of the 4-input demultiplexer*

The action of the four-input demultiplexer is illustrated by the waveforms in Figure 9.12.

Practical Exercise 9.3

To investigate the demultiplexer
For this exercise you will need the following components and equipment:

1 – 74LS139 ic (dual 1-of-4 demultiplexer)
1 – +5 V DC power supply
4 – LED (5 mm) and series resistor (270 Ω)

The pin connection diagram for the 74139 ic is shown in Figure 9.13.

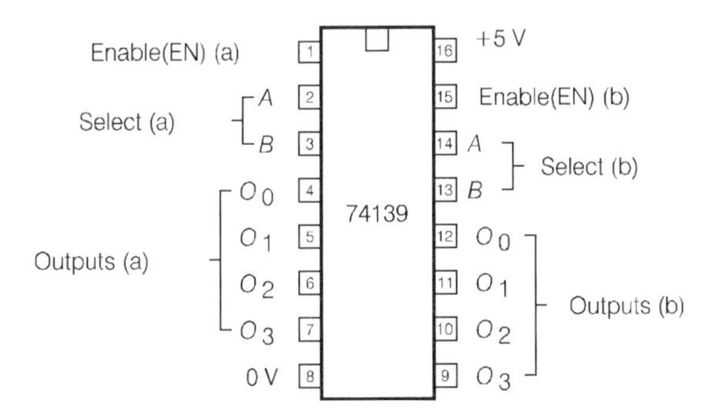

Figure 9.13 *The 74139 dual 1-of-4 demultiplexer: pin connections*

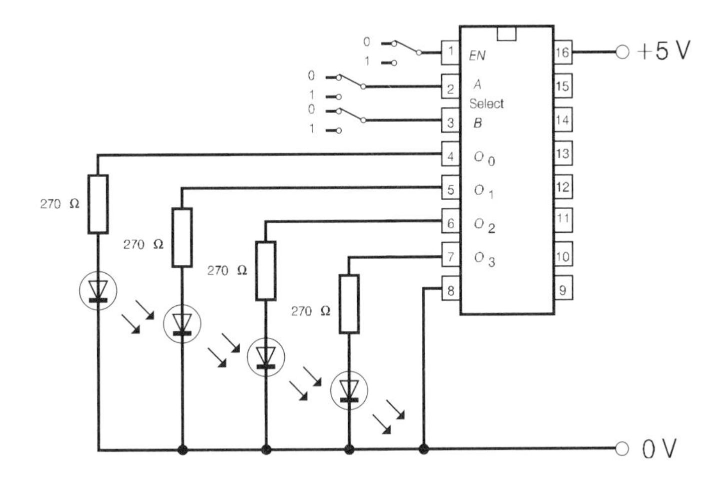

Figure 9.14 *The 74139 demultiplexer: circuit for Practical Exercise 9.3*

Continued on p. 188

Practical Exercise 9.3 (*Continued*)

For each circuit there is one enable (*EN*) and two channel select (*A*, *B*) inputs, and four (O_0 to O_3) outputs.

When the enable input is logic 1, all four outputs are logic 1, irrespective of the logic level of the select inputs.

When the enable input is logic 0, the selected output will be logic 0, with all other outputs remaining at logic 1.

The enable is used as the data input.

Procedure

1 Connect up the circuit of Figure 9.14.
2 Work through and complete the truth table shown in Figure 9.15.

EN	channel select		outputs			
	B	*A*	O_3	O_2	O_1	O_0
1	X	X				
0	0	0				
0	0	1				
0	1	0				
0	1	1				

Where an X is shown, it means that the logic level does NOT matter

Figure 9.15 *Truth table for Practical Exercise 9.3*

The data transmission system

The logical step now is to consider the combined effect of a multiplexer and a demultiplexer, which forms the basis of the next practical exercise.

Practical Exercise 9.4

To set up a data transmission system
For this exercise you will need the following components and equipment:

1 – 74LS153 ic (dual 4-to-1 multiplexer)
1 – 74LS139 ic (dual 1-of-4 demultiplexer)

1 – 74LS93 ic (4-bit binary counter)
1 – 74LS04 ic (hex inverter)
1 – +5 V DC power supply
1 – clock pulse generator (Figure 1.5)
4 – LED (5 mm) and series resistor (270 Ω)

Procedure

1 The outline block diagram of the system is given in Figure 9.16. For the pin connection diagrams you should refer as follows:

7493 ic – Figure 7.7
74153 ic – Figure 9.5
74139 ic – Figure 9.13
7404 ic – Figure 2.13(e)

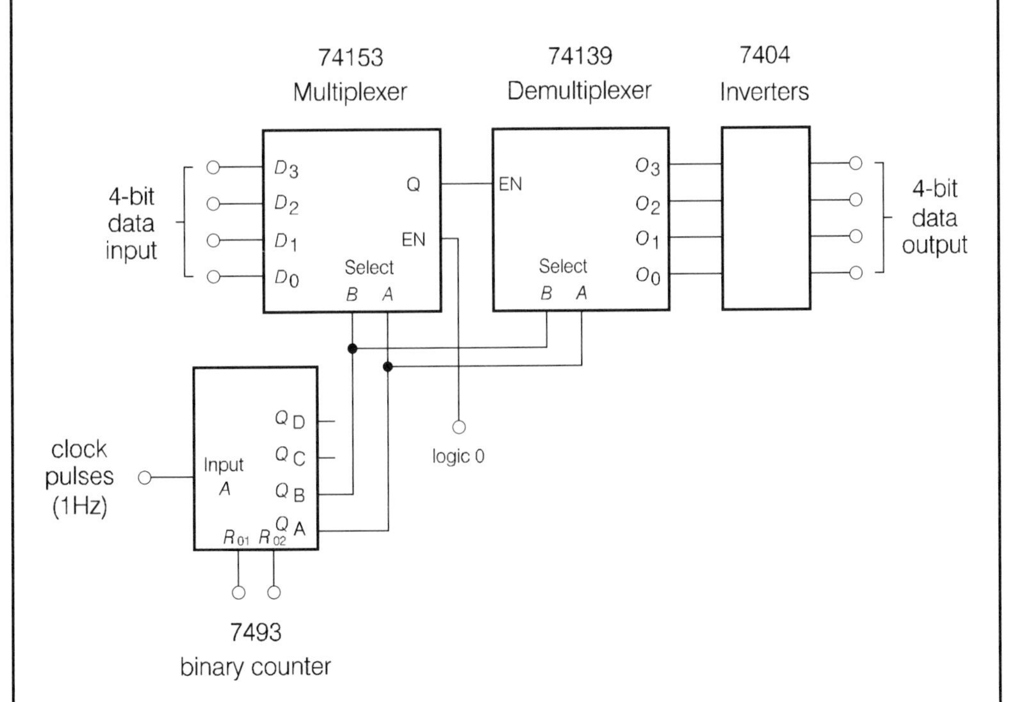

Figure 9.16 *The data transmission system: outline block diagram for Practical Exercise 9.4*

Conclusions

1 Since all the units used in this exercise have been met earlier, you should be able to explain the action of this system.

Continued on p. 190

Practical Exercise 9.4 *(Continued)*

2 Write an account of the operation of this system, taking as your basis the transfer of data $D_0D_1D_2D_3 = 1101$. Include in the write-up reference to all of the following:

(a) The purpose of the clock.
(b) The operation of the counter, starting from outputs $Q_DQ_CQ_BQ_A = 0000$.
(c) The inputs *AB* of both MUX and DMUX.
(d) The transfer of data from the input to output of MUX and the necessary logical state of the *EN* input for this transfer. Relate the logical state of this output to that of the counter.
(e) Transfer of data from input to output of DMUX and its relationship to the counter output.
(f) The need for the inverter of the DMUX output.

3 Copy out the following statements and fill in the spaces by using either the word *parallel* or *serial*, as appropriate:

(a) The input to the multiplexer is in _____ form.
(b) The output from the MUX is in _____ form.
(c) The input to the demultiplexer is in _____ form.
(d) The output from the DMUX is in _____ form.

Encoders

The encoder is a device which changes information from one form to another. For example, a particular requirement would be the need to convert decimal numbers from a keypad (telephone or computer) into binary coded decimal (BCD) form.

An encoder is normally to be found at the input of a system.

Important points

- An encoder has a number of input lines equal to the number of code combinations to be generated. Only one of these lines is activated at any one time.
- The number of output lines is equal to the number of bits in the code.

Taking as an example the 8421 BCD encoder, the truth table for which is shown in Figure 9.17, we see that to represent decimal numbers 0 to 9 requires a total of 10 input lines. There will thus be ten input lines (0 to 9) and four output lines (*ABCD*). This particular situation will be described in manufacturers' literature as a **10-line to 4-line priority encoder**.

Referring again to the keypad: for the above example the numerals 0 to 9 will be needed and are shown as switches S_0 to S_9 respectively (Figure 9.18).

INPUT	OUTPUTS			
Decimal	*D*	*C*	*B*	*A*
0	0	0	0	0
1	0	0	0	1
2	0	0	1	0
3	0	0	1	1
4	0	1	0	0
5	0	1	0	1
6	0	1	1	0
7	0	1	1	1
8	1	0	0	0
9	1	0	0	1

Figure 9.17 *The 8421 BCD code*

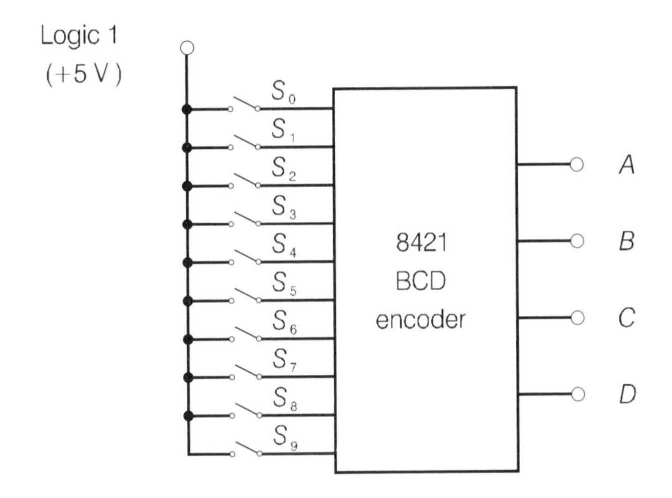

Figure 9.18 *The BCD encoder*

From Figures 9.17 and 9.18, we can deduce that output D will be logic 1 if *either* input S_8 *or* input S_9 is closed, which gives the logic statement

$$D = S_8 + S_9$$

Similarly it can be shown that

$$C = S_4 + S_5 + S_6 + S_7$$

$$B = S_2 + S_3 + S_6 + S_7$$

$$A = S_1 + S_3 + S_5 + S_7 + S_9$$

This leads to the logic circuit diagram of Figure 9.19, which could of course be made up using basic logic gates. The 74147 ic will be used instead, in the next practical exercise.

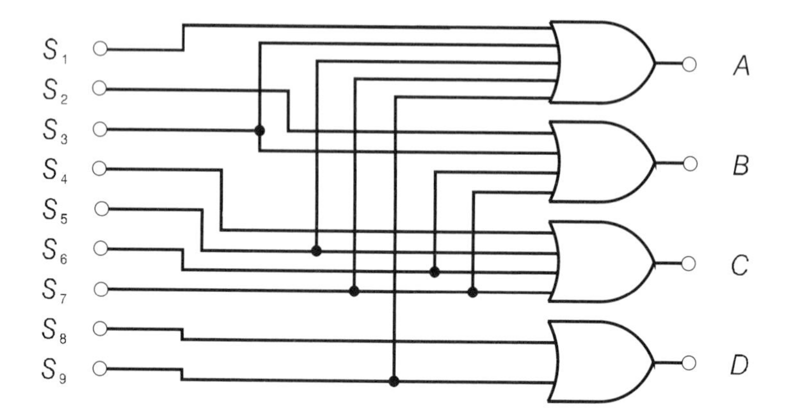

Figure 9.19 *Implementation of encoder example*

Practical Exercise 9.5

To investigate the encoder
For this exercise you will need the following components and equipment:

1 – 74LS147 ic (10-line to 4-line priority encoder)
1 – +5 V DC power supply
4 – LED (5 mm) and series resistor (270 Ω)

The pin connection diagram of the 74147 ic is shown in Figure 9.20. Note that the inputs and outputs are **active low**. In other words, a logic 0 at the input is necessary to activate the corresponding output which will then also be logic 0.

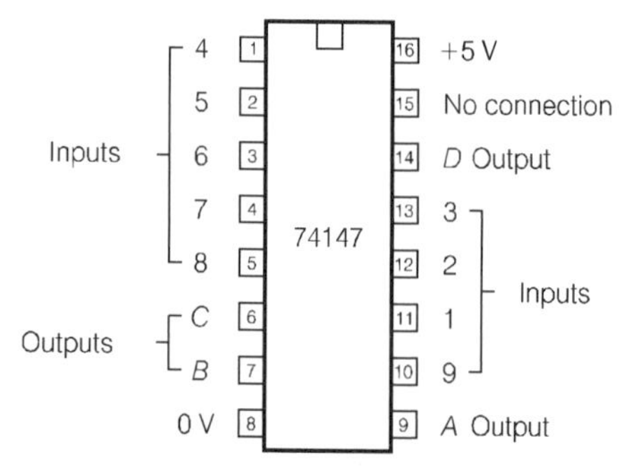

Figure 9.20 *The 74147 10-line to 4-line encoder: pin connections*

Procedure

1 Connect up the circuit of Figure 9.21.

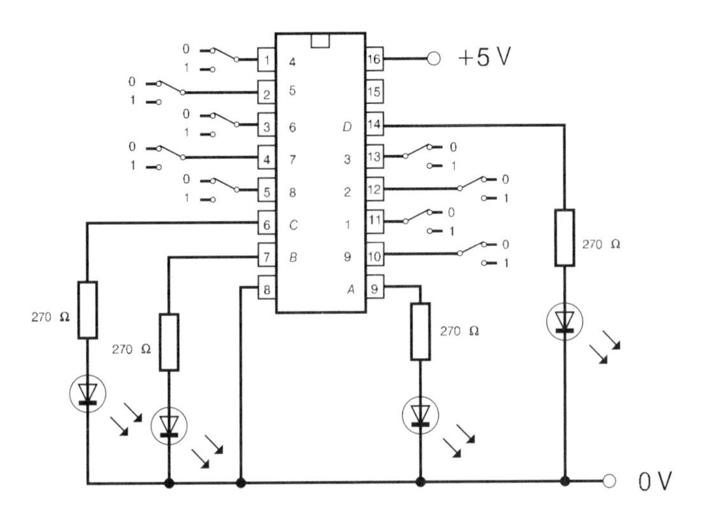

Figure 9.21 *The 74147 10-line to 4-line encoder: circuit for Practical Exercise 9.5*

2 Work through and confirm the truth table of Figure 9.22.

INPUTS									OUTPUTS			
1	2	3	4	5	6	7	8	9	*D*	*C*	*B*	*A*
1	1	1	1	1	1	1	1	1	1	1	1	1
X	X	X	X	X	X	X	X	0	0	1	1	0
X	X	X	X	X	X	X	0	1				
X	X	X	X	X	X	0	1	1				
X	X	X	X	X	0	1	1	1				
X	X	X	X	0	1	1	1	1				
X	X	X	0	1	1	1	1	1				
X	X	0	1	1	1	1	1	1				
X	0	1	1	1	1	1	1	1				
0	1	1	1	1	1	1	1	1				

Where an X is shown, it means that the logic level does NOT matter

Figure 9.22 *Truth table for Practical Exercise 9.5*

Continued on p. 194

Practical Exercise 9.5 (*Continued*)

Conclusion

1 Draw up the truth table for the following encoders:
   (a) 16-line to 4-line
   (b) 8-line to 3-line

Decoders

Like the encoder, this device changes information from one form to another, but performs the opposite function to that of an encoder. It is used at the output of a system.

Important points

- The number of input lines equals the number of bits in the code.
- The number of output lines equals the number of code combinations.
- For a given binary input, one of the output lines will go to logic 1, with all other output lines remaining at logic 0.

The decoder block diagram is shown in Figure 9.23, which you should recognize as the opposite to the encoder of Figure 9.18. It is known as a **4-line to 10-line decoder**. The 4-bit binary code is shown in Figure 9.24 and the input and output line arrangement in Figure 9.25.

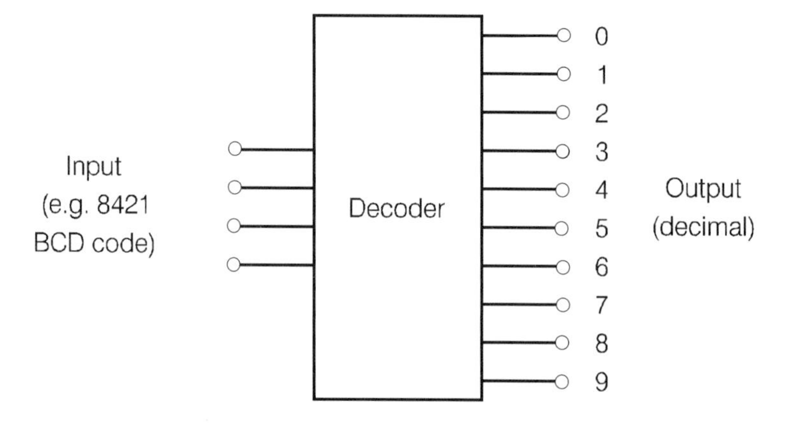

Figure 9.23 *The decoder block diagram*

D	C	B	A	Output = 1
0	0	0	0	Z_0
0	0	0	1	Z_1
0	0	1	0	Z_2
0	0	1	1	Z_3
0	1	0	0	Z_4
0	1	0	1	Z_5
0	1	1	0	Z_6
0	1	1	1	Z_7
1	0	0	0	Z_8
1	0	0	1	Z_9
1	0	1	0	Z_{10}
1	0	1	1	Z_{11}
1	1	0	0	Z_{12}
1	1	0	1	Z_{13}
1	1	1	0	Z_{14}
1	1	1	1	Z_{15}

Figure 9.24 *The 4-bit binary decoding table*

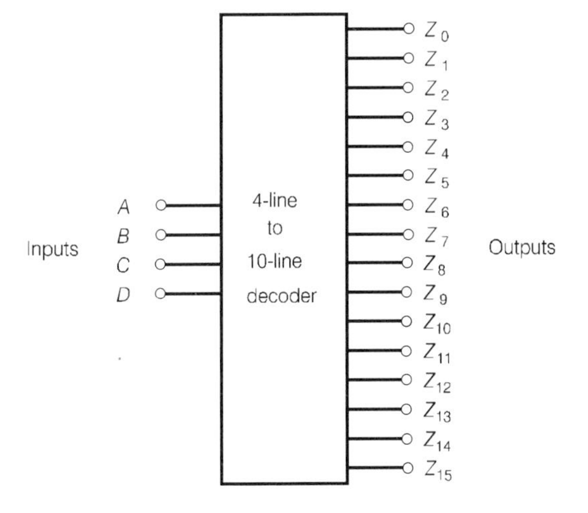

Figure 9.25 *The 4-line to 10-line decoder block diagram*

The logic statements for the various outputs can be written as follows:

$$Z_0 = \overline{A} \cdot \overline{B} \cdot \overline{C} \cdot \overline{D}$$

$$Z_1 = A \cdot \overline{B} \cdot \overline{C} \cdot \overline{D}$$

and so on.

The 4-line to 10-line decoder finds application in converting from BCD to decimal, as the next practical exercise will show.

Question

9.2 Draw up the truth table and write down the logic statements for the 3-line to 8-line decoder.

Draw the logic diagram and, with the usual availability of AND, OR and NOT gates, estimate the overall requirements for implementation.

Compare the cost of the above arrangement with that of the equivalent single ic.

Practical Exercise 9.6

To investigate the decoder

For this exercise you will need the following components and equipment:

1 – 74LS42 ic (BCD to decimal decoder)
1 – +5 V DC power supply
10 – LED (5 mm) and series resistor (270 Ω)

Procedure

1 Connect up the circuit of Figure 9.26. The pin connection diagram of the 7442 ic is shown in Figure 9.27.
2 Work through and complete the truth table of Figure 9.28. You should notice that, for a particular binary input, the corresponding decimal output is logic 0 with the

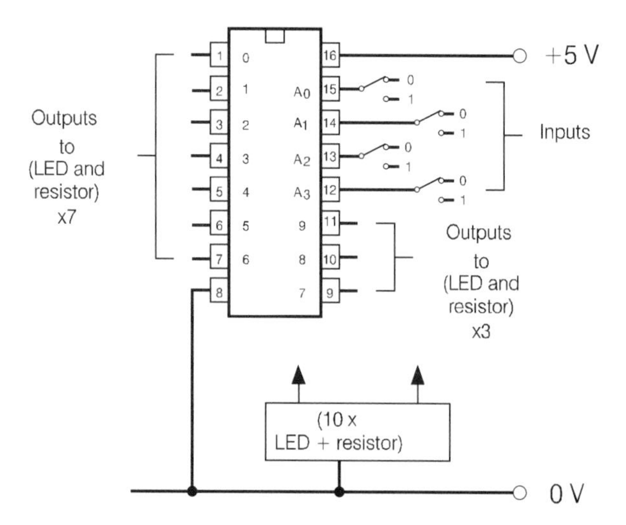

Figure 9.26 *The 7442 BCD to decimal decoder: circuit for Practical Exercise 9.6*

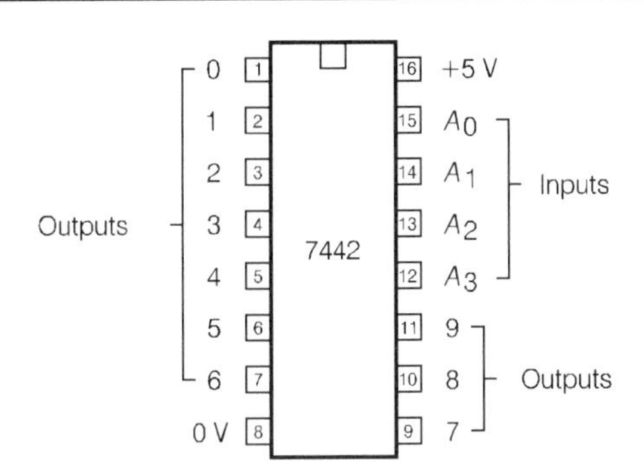

Figure 9.27 *The 7442 BCD to decimal decoder: pin connections*

BCD Inputs				DECIMAL Outputs									
D	*C*	*B*	*A*	9	8	7	6	5	4	3	2	1	0
0	0	0	0	1	1	1	1	1	1	1	1	1	0
0	0	0	1	1	1	1	1	1	1	1	1	0	1
0	0	1	0										
0	0	1	1										
0	1	0	0										
0	1	0	1										
0	1	1	0										
0	1	1	1										
1	0	0	0										
1	0	0	1										

Figure 9.28 *Truth table for Practical Exercise 9.6*

LED being unlit. Thus the outputs are **active low**, which accounts for the style of the truth table.

Conclusion

How should the LED and resistor be connected in order for the LED to be lit when the output is low?

Hint. Look ahead to Chapter 10!

Code converters

As the name suggests, in the code converter the digital input information is converted from one code to another. It follows that encoders and decoders come into this category.

Another typical example of the decoder is the 7447 BCD to seven-segment display (dealt with in chapter 10).

A code converter that converts decimal (and hexadecimal) numbers into coded form is thus an encoder, the 74147 (dealt with earlier) and 74148 (8-to-3) being examples.

Adders

A binary adder is a device that will add together two binary numbers.

From your knowledge of decimal addition you will recall that when numbers are added together, for a particular column there will always be **a sum** (naturally!) and sometimes **a carry**.

Examples

```
  5
 +4
 ——
  9 (the sum)
 ——

  5
 +7
 ——
  2 (the sum)
 ——
 1   (the carry)
```

There are two types of binary adder:

(a) The half-adder, which will provide a sum and a carry, but cannot take into account any carry from a previous stage.
(b) The full-adder, which takes into account a previous carry.

It would be as well at this point to refresh our memory on the rules of binary addition, which are crucial to the understanding of half- and full-adders. These rules are:

$0 + 0 = 0$ carry 0

$0 + 1 = 1$ carry 0

$1 + 0 = 1$ carry 0

$1 + 1 = 0$ carry 1

Some worked examples of binary addition are given in Figure 9.29.

Figure 9.30 shows these rules in the form of a truth table, which is the truth table for the half-adder. In addition, the block diagram is shown in Figure 9.31. From the truth table, the logic statements for the half-adder are given by

$$S = A \cdot \overline{B} + \overline{A} \cdot B$$

$$C_o = A \cdot B$$

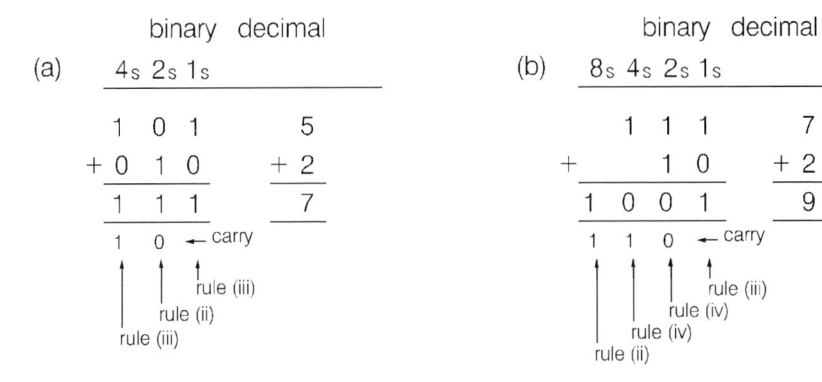

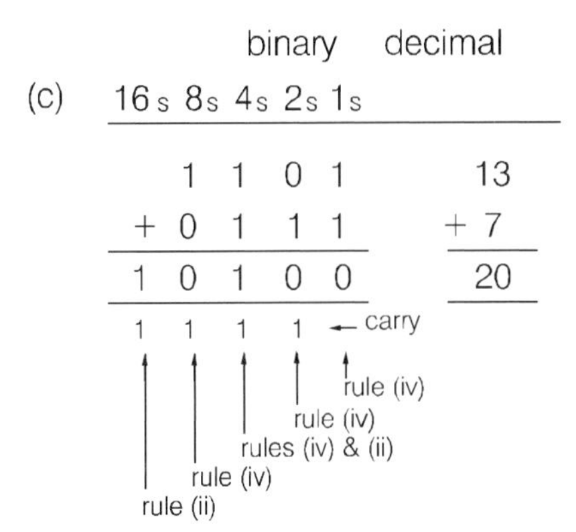

Figure 9.29 *Worked examples of binary addition*

INPUTS		OUTPUTS	
B	A	Sum (S)	Carry (C_o)
0	0	0	0
0	1	1	0
1	0	1	0
1	1	0	1

Figure 9.30 *The half-adder truth table*

The implementation of these statements using the basic gates is shown in Figure 9.32.

Notice from the truth table in Figure 9.30 that the sum column is that of an exclusive OR gate. This makes the implementation somewhat easier, as shown in Figure 9.33.

The full-adder block diagram is shown in Figure 9.34 and the truth table in Figure 9.35. Notice that the truth table is showing all the possible combinations of inputs A and B and the carry C_o.

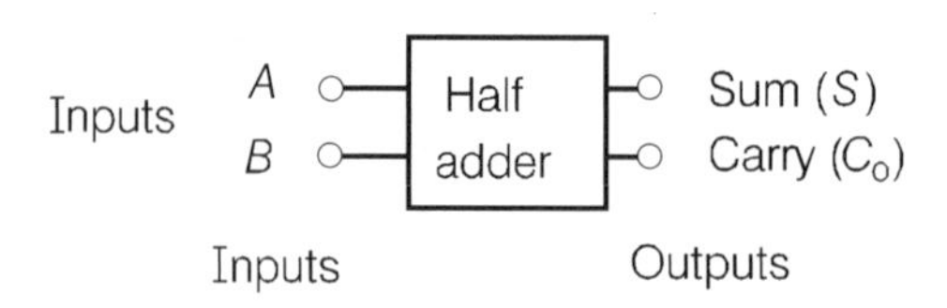

Figure 9.31 *The half-adder block diagram*

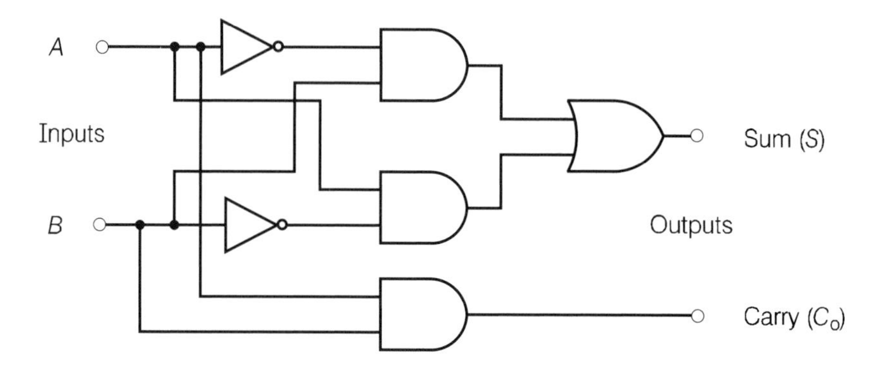

Figure 9.32 *The half-adder: implementation using basic gates*

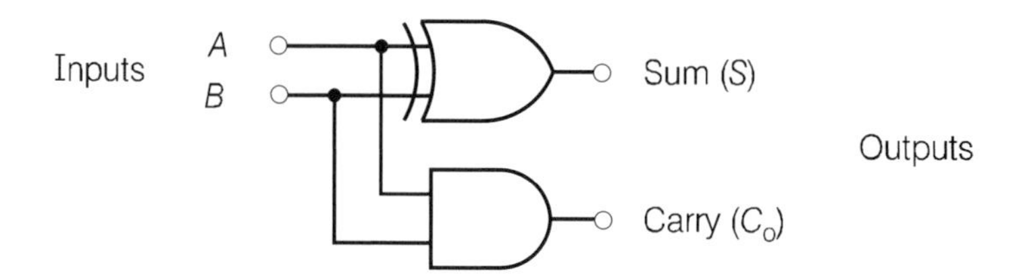

Figure 9.33 *The half-adder: simplified logic diagram*

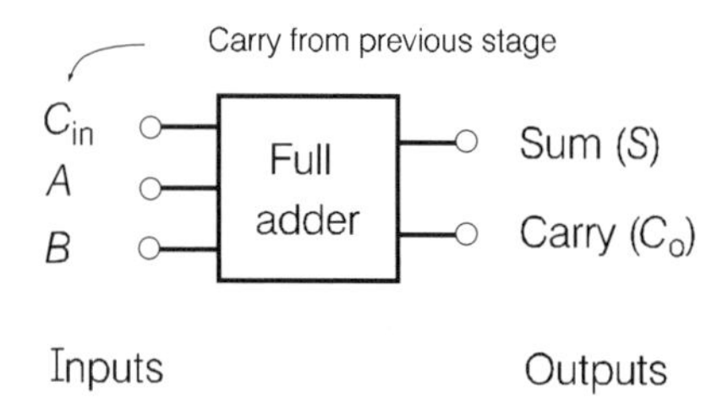

Figure 9.34 *The full-adder block diagram*

The logic statements are given by

$$S = A \cdot \overline{B} \cdot \overline{C}_{in} + \overline{A} \cdot B \cdot \overline{C}_{in} + \overline{A} \cdot \overline{B} \cdot C_{in} + A \cdot B \cdot C_{in}$$

$$C_o = A \cdot B \cdot \overline{C}_{in} + A \cdot \overline{B} \cdot C_{in} + \overline{A} \cdot B \cdot C_{in} + A \cdot B \cdot C_{in}$$

$$= A \cdot C_{in} + B \cdot C_{in} + A \cdot B$$

(Karnaugh mapping will prove this simplification for C_o.)

INPUTS			OUTPUTS	
C_{in}	B	A	Sum (S)	Carry (C_o)
0	0	0	0	0
0	0	1	1	0
0	1	0	1	0
0	1	1	0	1
1	0	0	1	0
1	0	1	0	1
1	1	0	0	1
1	1	1	1	1

Figure 9.35 *The full-adder truth table*

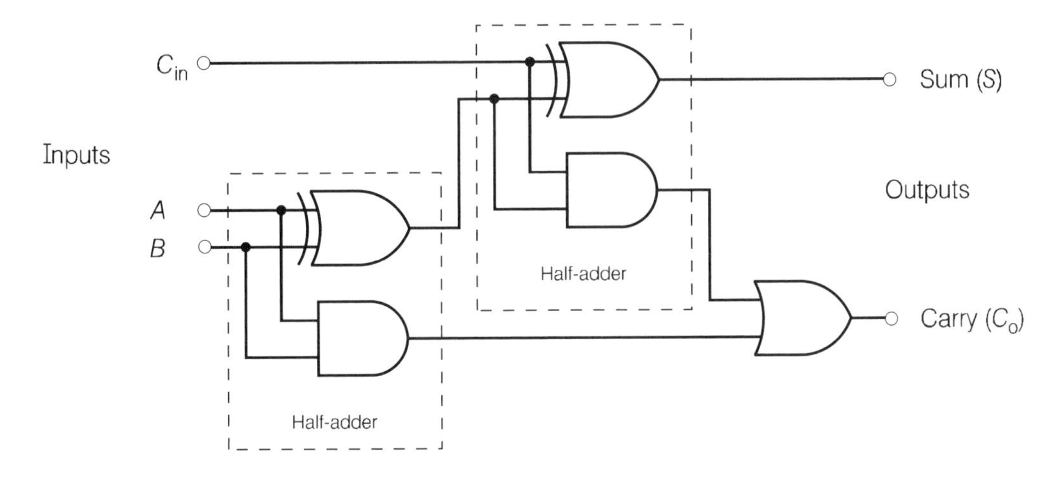

Figure 9.36 *The full-adder logic diagram*

The full-adder can be made up using two half-adders and an OR gate, as shown in Figure 9.36. The modified block diagram is shown in Figure 9.37.

The 4-bit parallel adder

When adding together two n-bit numbers, it is preferable, from the speed of operation point of view, to use the parallel rather than series (serial) arrangement. This is shown by the diagram in Figure 9.38.

Two 4-bit numbers, given by $A_1A_2A_3A_4$ and $B_1B_2B_3B_4$, are to be added together. The least significant bit (2^0) for each input is A_1 and B_1 respectively. It follows therefore, that in practice, the numbers will normally be presented in the reverse order, as $A_4A_3A_2A_1$ and $B_4B_3B_2B_1$ respectively.

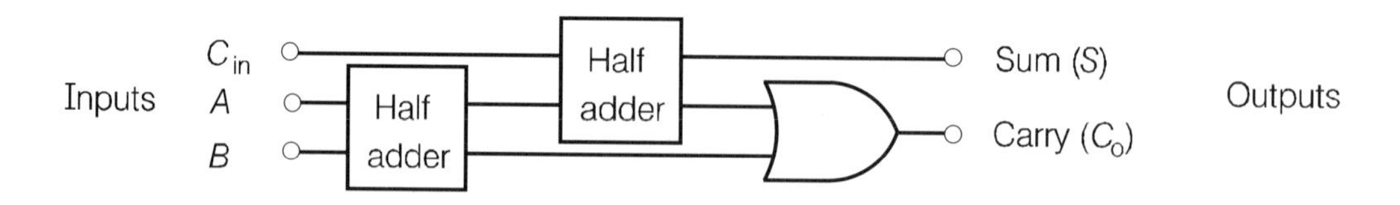

Figure 9.37 *The full-adder modified block diagram*

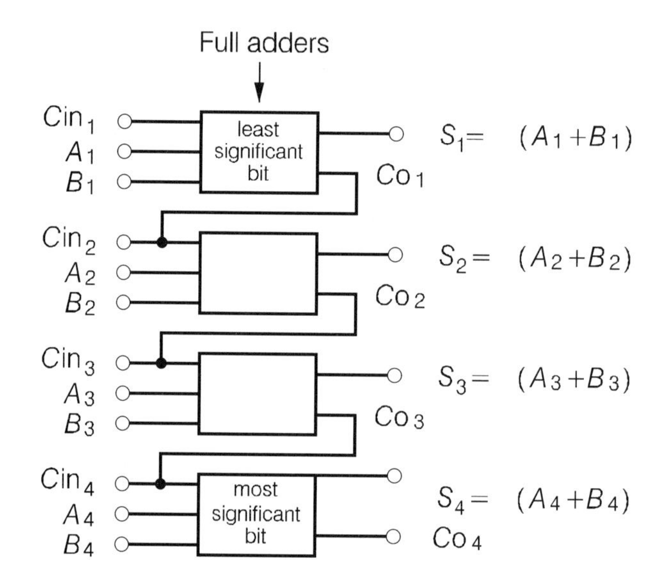

Figure 9.38 *The 4-bit parallel adder*

There will be one **sum** output (S) and one **carry** output (C_o) for each of the four bits. S_1 is the least significant output bit and the carry-outs C_{01}, C_{02} and C_{03} are each connected to the carry-in of their following stage. The fourth stage provides the resultant output carry C_{04}.

If there is no carry-in Cin_1, this will be connected to logic 0.

Example

Add together the 4-bit numbers $A_4A_3A_2A_1 = 1010$ (decimal 10) and $B_4B_3B_2B_1 = 1110$ (decimal 14).

Figure 9.39 shows the situation, with the resultant sum being $A_5A_4A_3A_2A_1 = 11000$ (decimal 24) as expected.

All the bits in the number are added **at the same time** (in parallel) as opposed to one after the other (in series). This makes for a faster addition time.

Computer systems requiring addition of 8-bit and 16-bit numbers will use the above principle involving full-adder stages. An 8-bit addition is shown in Figure 9.40.

Four full adders can be combined within one ic (as for example, the 74283), which forms the basis of the next Practical Exercise.

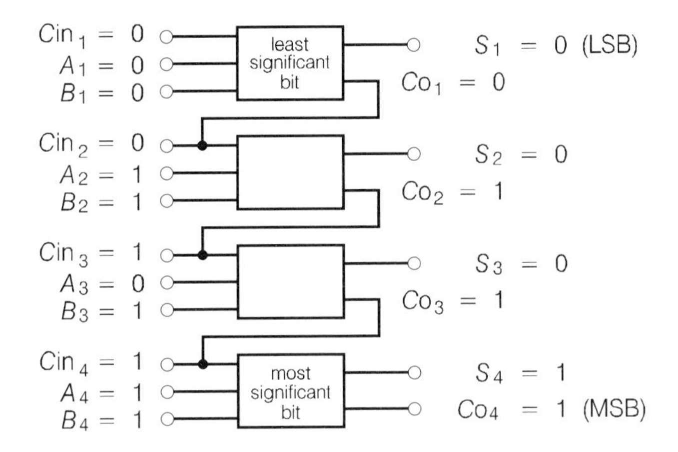

Figure 9.39 *The 4-bit parallel adder: diagram for the addition of the numbers 1010 and 1110*

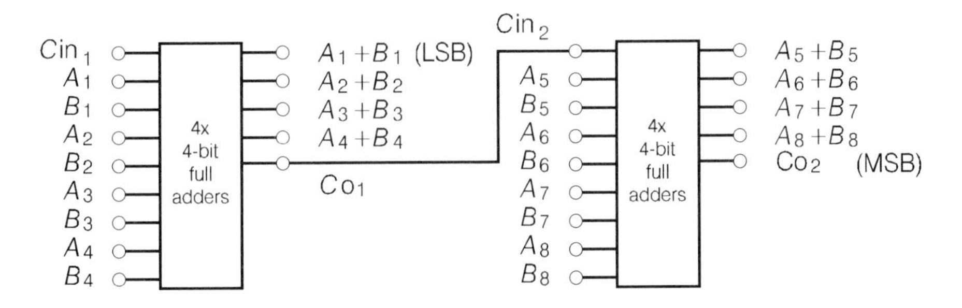

(a) Outline diagram

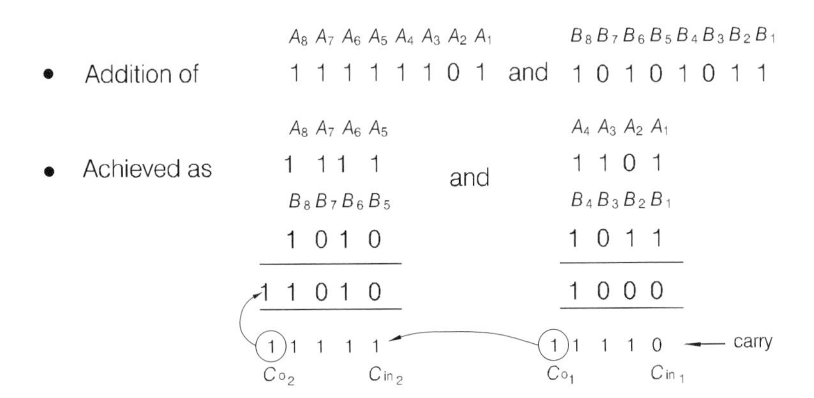

- Giving the answer 110101000

(b) Numerical example

Figure 9.40 *8-bit addition: outline diagram and numerical example*

Question

9.4 Suggest, by means of an outline diagram, how two 16-bit numbers could be added together using 4-bit parallel adders.

Practical Exercise 9.7

To investigate the full adder
For this exercise you will need the following components and equipment:

1 – 74LS283 ic (4-bit binary full-adder)
1 – +5 V DC power supply
5 – LED (5 mm) and resistor (270 Ω)

Procedure (a)

1 Connect up the circuit shown in Figure 9.41. The pin connection diagram for the 74283 ic is shown in Figure 9.42.
Notice that, for this first part of the exercise, all four *A* inputs are connected together, as are all the four *B* inputs. With this arrangement, the addition is merely of two binary numbers

2 Apply the appropriate logic levels to inputs *A* and *B* respectively in order to verify the rules of binary addition given above.

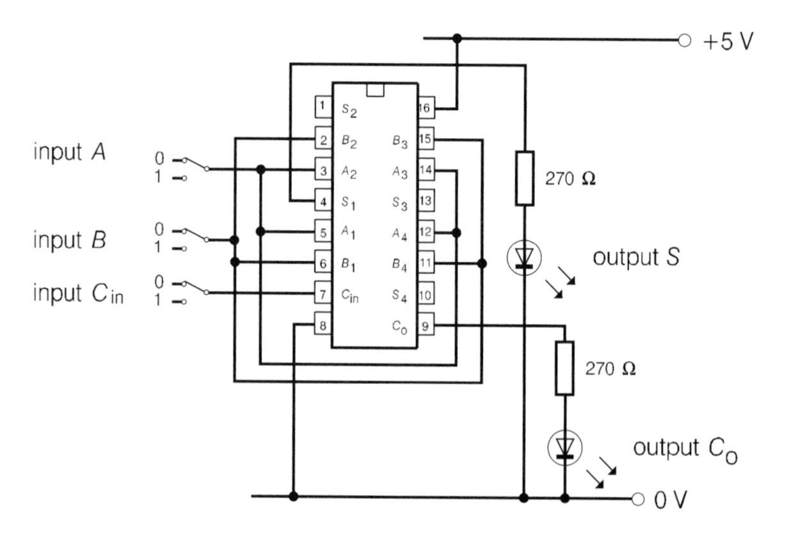

Figure 9.41 *The 74283 as a 2-bit binary full-adder: circuit for Practical Exercise 9.7(a)*

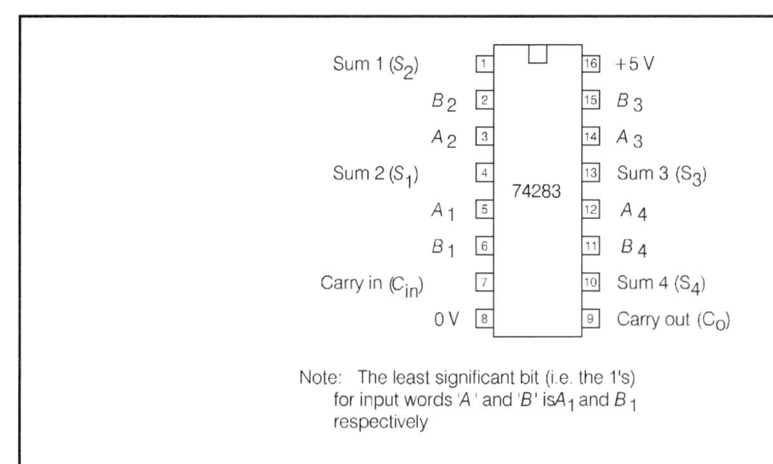

Figure 9.42 *The 74283 4-bit binary full-adder: pin connections*

Procedure (b)

1. The previous circuit should now be modified to that shown in Figure 9.43.
2. It will probably be more convenient if one input word (say, A) is left set up and the addition performed by setting the other as required. Thus.
3. Set input $A_4A_3A_2A_1$ to 0101 (decimal 5) and let this remain for the duration of this part of the exercise.

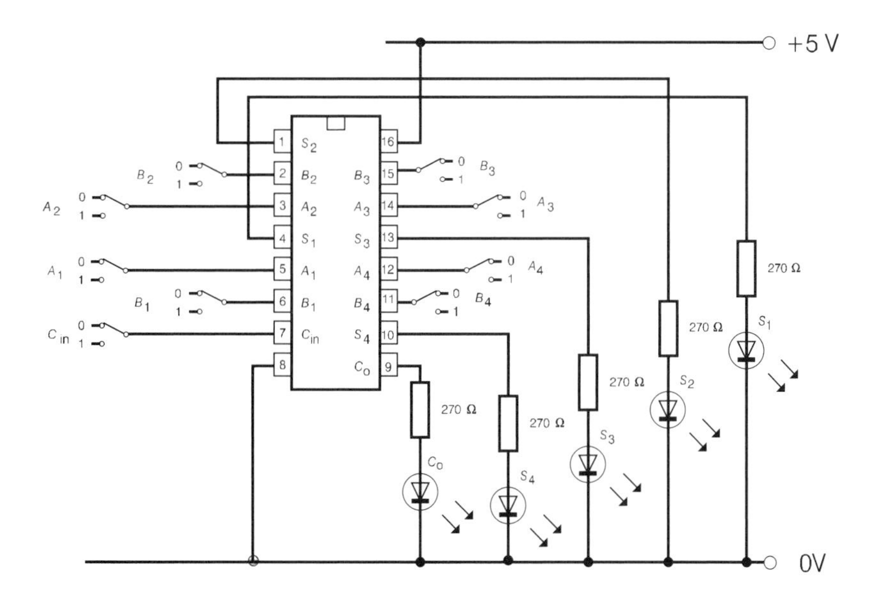

Figure 9.43 *The 74283 4-bit binary full-adder: circuit for Practical Exercise 9.7(b)*

Continued on p. 206

Practical Exercise 9.7 (*Continued*)

		8s	4s	2s	1s	decimal
(a)	A	0	1	0	1	5
	B	0	0	0	1	+1
						6
(b)	A	0	1	0	1	
	B	0	0	1	0	
(c)	A	0	1	0	1	
	B	0	1	0	1	
(d)	A	0	1	0	1	
	B	0	1	0	1	
	C_{in}				1	
(e)	A	0	1	0	1	
	B	1	1	1	1	
	C_{in}				1	

Figure 9.44 *Examples for Practical Exercise 9.7(b)*

4 Set input $B_4B_3B_2B_1$ to the levels shown in parts (a) to (e) of Figure 9.44. Obtain the answers practically and theoretically, checking that the binary output agrees with its decimal equivalent.

Question

9.3 Add together the following binary numbers, checking your answers with the decimal equivalents:

(a) 1000 + 1000
(b) 1001 + 0111
(c) 1010 + 0101 + 1001
(d) 1110 + 0101 + 1101
(e) 10100111 + 00110101

10 Display devices

In today's digital world there is an increasing need for information to be displayed in a form that is instantly recognizable and readable. This usually means 'display by digits', the output being time of day, money spent in a supermarket, petrol taken from a pump, temperature in a particular place and so on.

The general name given to the area of work in which electronics and optics combine to enable this information to be conveyed is called (naturally enough) **optoelectronics**.

Within this area, various devices are used to display numbers, letters and other types of message.

The light emitting diode (LED)

The LED is a pn junction diode made from semiconductor materials, usually in the form of gallium phosphide or gallium arsenide phosphide.

Its function is to emit light of a certain colour (i.e. wavelength) when forward biased. The amount of forward current (I_f) will determine the brightness, but this current must be limited to a safe value, otherwise the diode will be destroyed. A **current limiting resistor** (R_s) is used and the circuit is shown in Figure 10.1. LEDs are usually used to indicate the presence or absence of logic levels, which implies a supply voltage (V_s) of +5 V for TTL logic.

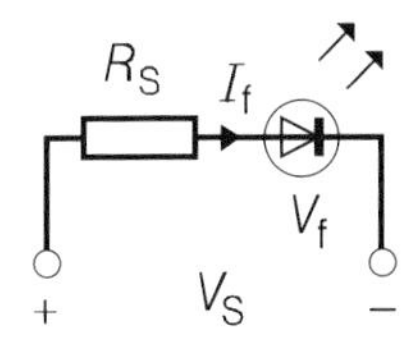

Figure 10.1 *Biasing the LED*

A typical forward voltage drop across the LED is 2 V for a typical forward current of 10 mA, although with the variety of devices available, other values can be used. However we shall assume these values apply for this example in calculating the necessary value of the limiting resistor.

From the circuit in Figure 10.1 we can write

$$V_s = I_f R_s + V_f$$

giving

$$R_s = \frac{(V_s - V_f)}{I_f}$$

from which, if $V_s = 5$ V, $V_f = 2$ V and $I_f = 10$ mA,

$$R_s = \frac{(5-2)\text{ V}}{10\text{ mA}}$$

$$= 300\ \Omega \text{ (the choice of preferred value is 270 }\Omega\text{ or 330 }\Omega\text{)}$$

Important points

- Correct polarity is essential! The diode can be destroyed, not only by excessive forward current, but also by excessive reverse voltage. A typical maximum safe reverse voltage is −5 V.
- The circuit is not restricted to the use of a +5 V supply provided that the rules regarding current limiting are followed.
- LEDs use less current than filament lamps and are more reliable. The single colours available are red, green, yellow and orange.
- Other available types are **bicolour** (red and green) where two LEDs are connected in parallel but in opposite directions – called inverse parallel – and **tricolour** (red, yellow and green) with a common cathode and same supply voltage polarity, where the colour will depend on the relative currents through the diodes.

Practical Exercise 10.1

To investigate the LED
For this exercise you will need the following components and equipment:

1 – LED (5 mm red) and 270 Ω resistor
1 – 0 to +5 V (i.e. variable) DC power supply
1 – 0–50 mA DC meter
1 – 0–10 V DC meter

Procedure

1 Connect up the circuit of Figure 10.2. The LED pin connections are shown in Figure 1.6.
2 Vary the DC power supply voltage from 0 to +5 V in suitable steps and note for each step the ammeter and voltmeter readings. Redraw and complete the table shown in Figure 10.3.

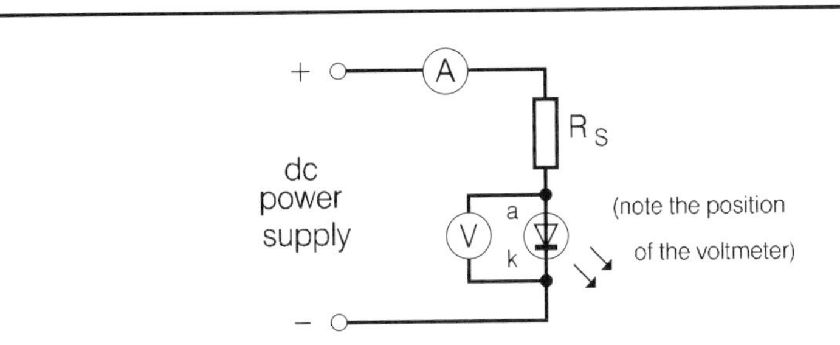

Figure 10.2 *Characteristics of the LED: circuit for Practical Exercise 10.1*

Supply voltage (V)	0	1	2	3	4	5
current through LED (mA)						
voltage across LED (V)						

Figure 10.3 *Table of values for Practical Exercise 10.1*

3 Plot a graph of current through the diode (vertically) against voltage across the diode.

Conclusions

1 From the graph, what comment can you make regarding the voltage across the diode once the current has reached a certain value?
What is this voltage value and at what value of current does it occur?

type → of LED	3 mm diameter	5 mm diameter (standard)	5 mm diameter (low current)	high brightness (standard)
colour $I_{f\ typical}$ $I_{f\ max}$ V_f $V_{r\ max}$				

Figure 10.4 *Table of data for Practical Exercise 10.1*

Continued on p. 210

Practical Exercise 10.1 *(Continued)*

2 Consult the manufacturers' data or suppliers' catalogues in order to compile a table of values, using Figure 10.4 as a guide.
3 Give an example where you think the tricolour LED (red, yellow, green) might be used.

Questions

10.1 Suppose you find that the only LED available for Practical Exercise 10.1 has a voltage drop of 1.8 V at a current of 20 mA. What would be a suitable value for the series resistor?
10.2 An LED (2 V at 10 mA) is to be used with a supply voltage of 9 V. Calculate the necessary value of the series resistor.

LEDs as indicators

The LED has been used with a large number of practical exercises within this book to indicate the **presence of logic 1** (+5 V), as in Figure 10.5(a). It can also be used to indicate the **absence of logic 1**, that is, the presence of logic 0 (0 V). This is shown in Figure 10.5(b).

Note the use of the terms current **sourcing** and **sinking**.

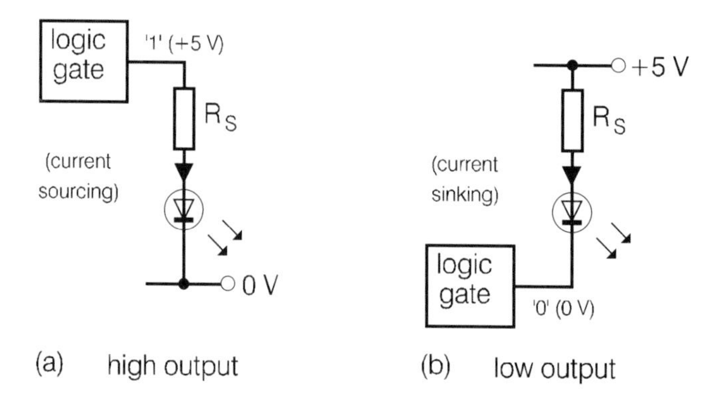

Figure 10.5 *LED indicators*

The seven-segment display

The diagrammatic arrangement is shown in Figure 10.6(a). All the anodes are connected together (called **common anode**) and to the +5 V supply. A particular segment will be lit when its cathode is connected, **through a current limiting resistor**, to 0 V.

Identification of the segment letters is given in Figure 10.6(b). Displays are also available with the common cathode connection.

Numbers 0 to 9 inclusive will be indicated as shown in Figure 10.7. A decimal point can be included, which may be to the right or the left of the display.

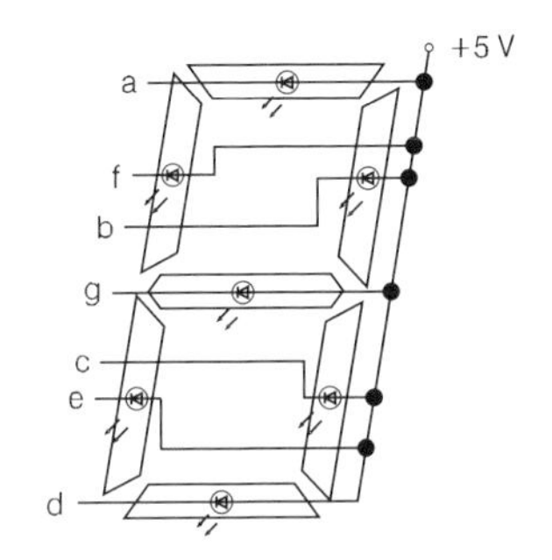

(a) common anode

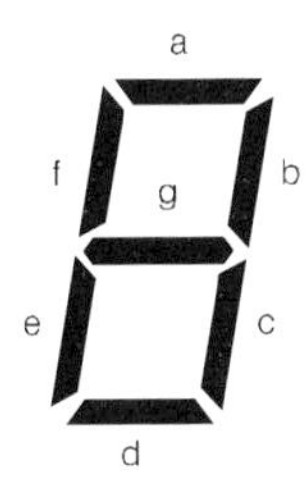

(b) segment letters

Figure 10.6 *The seven-segment display*

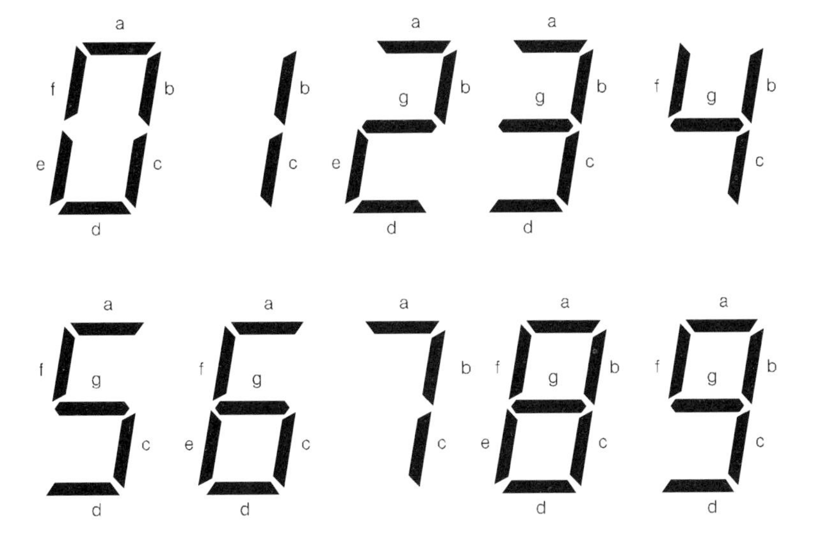

Figure 10.7 *The seven-segment display showing numbers 0 to 9*

The seven-segment package is available in various sizes of display height, ranging from 0.3 inch to 2.24 inches, with red being the favourite colour.

Important points

- A series resistor must always be used for each segment, including the decimal point if available, otherwise damage to that segment can result.
- Since the usual current for a sensible degree of brightness is of the order of 10 mA, it follows that the current with all segments lit (the number 8) will be 80 mA. Hence the use of a several-figure display will require a substantial continuous current, which tends to rule it out for battery-driven units.
- In order to provide the 7-bit input signal, a decoder/driver circuit is needed.

Practical Exercise 10.2

To investigate the seven-segment display
For this exercise you will need the following components and equipment:

1 – 0.3 inch seven-segment common anode display
8 – resistor (270 Ω)
1 – +5 V DC power supply
1 – 0–100 mA DC meter

Procedure

1. Connect up the circuit of Figure 10.8. The pin connections for the seven-segment display are shown in Figure 10.9.
2. By connecting the appropriate resistors to 0 V, check that numbers 0 to 9 and the decimal point can be displayed. If this does not happen, it is likely that a segment is damaged!
3. Connect the resistors so that the following segments are lit and note the shape of the display:
   (a) d, e and g
   (b) c, d and g
   (c) b, f and g
   (d) a, d, f and g
   (e) d, e, f and g.
4. Measure the current taken from the supply when all segments and the decimal point are being displayed.

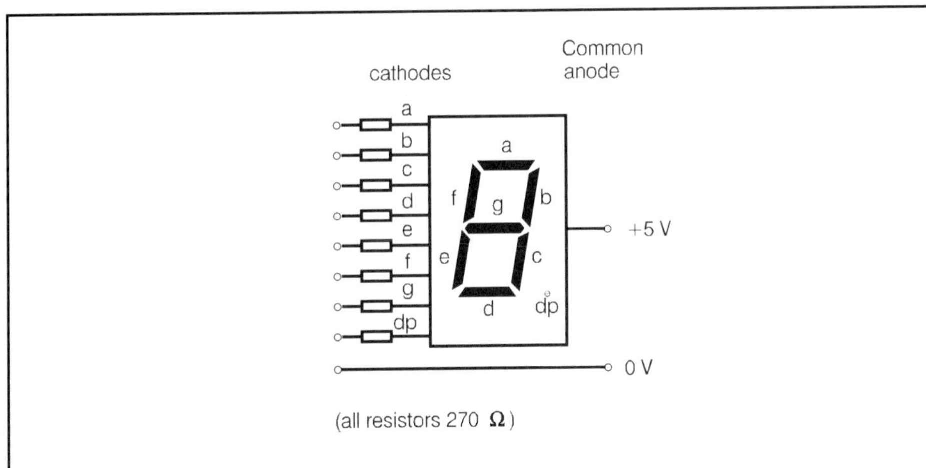

Figure 10.8 *The seven-segment display: circuit for Practical Exercise 10.2*

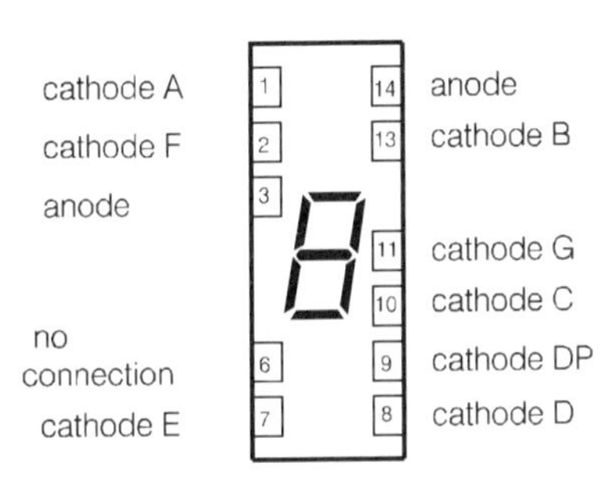

Figure 10.9 *The seven-segment display (common anode): pin connections*

Conclusion

State the logic level necessary at the input of the seven-segment display in order for the segments to be lit.

The decoder/driver

The 4-line to 7-line decoder accepts a binary coded decimal (BCD) number at the four inputs A, B, C and D and provides the 7-line signal for the display at outputs a to g. The block diagram for such a device is similar to that met earlier in Chapter 9.

A suitable decoder/driver for the common anode display is the 7447A integrated circuit, which has active-low open-collector outputs. Remember, **active-low** means that for a particular segment to be lit, the corresponding output from the driver must be logic 0.

Practical Exercise 10.3

To investigate the decoder/driver with seven-segment display
For this exercise you will need the following components and equipment:

1 – 7447A ic (BCD to seven-segment decoder)
1 – 0.3 inch seven-segment common anode display
7 – resistor (270 Ω)
1 – +5 V DC power supply

Procedure

1 Connect up the circuit shown in Figure 10.10. The pin connections for the 7447A ic are shown in Figure 10.11 and those for the seven-segment display in Figure 10.9.

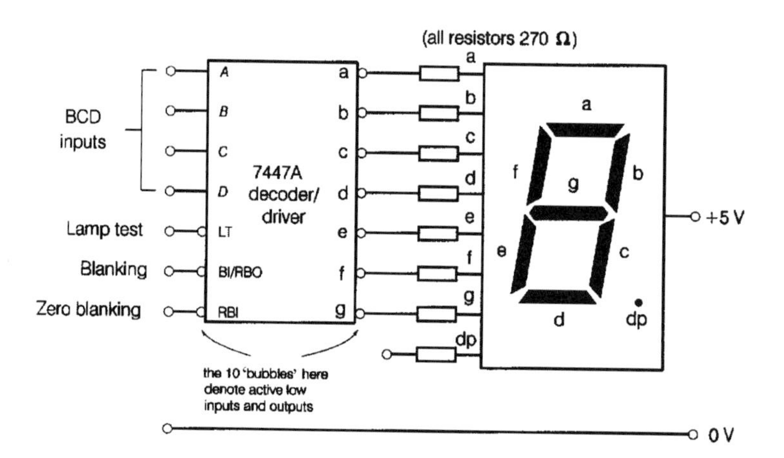

Figure 10.10 *The 7447A decoder driver with seven-segment display: circuit for Practical Exercise 10.3*

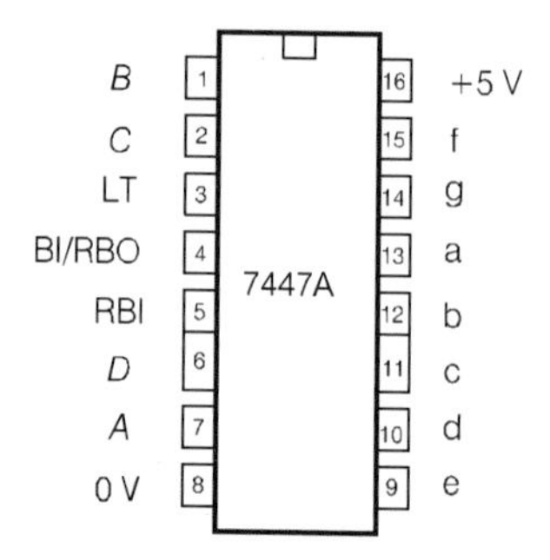

Figure 10.11 *The 7447A decoder/driver: pin connections*

2 Draw up the outline truth table shown in Figure 10.12. Remember, input *A* represents the least significant bit (LSB).
3 Work through and complete this table by applying the inputs shown. Write 1 in the table when a segment is lit and 0 when it is not.

Decimal/ function	INPUTS							OUTPUTS							Display
	LT	RBI	*D*	*C*	*B*	*A*	BI/RBO	a	b	c	d	e	f	g	
0	1	1	0	0	0	0	1	1	1	1	1	1	1	0	0
1	1	X	0	0	0	1	1								
2	1	X	0	0	1	0	1								
3	1	X	0	0	1	1	1								
4	1	X	0	1	0	0	1								
5	1	X	0	1	0	1	1								
6	1	X	0	1	1	0	1								
7	1	X	0	1	1	1	1								
8	1	X	1	0	0	0	1								
9	1	X	1	0	0	1	1								
10	1	X	1	0	1	0	1								
11	1	X	1	0	1	1	1								
12	1	X	1	1	0	0	1								
13	1	X	1	1	0	1	1								
14	1	X	1	1	1	0	1								
15	1	X	1	1	1	1	1								
BI	X	X	X	X	X	X	0								
RBI	1	0	0	0	0	0	0								
LT	0	X	X	X	X	X	1								

X = 'does not matter' 1 = 'lamp lit'

Figure 10.12 *Outline truth table for Practical Exercise 10.3*

Conclusions

In working through this exercise you will have observed the function of the other inputs to the 7447 ic. In addition, you will have seen the display following the application of the so-called **illegal combination** of inputs, namely decimal 10 to 15. The following four incomplete statements represent a summary of these observations. They should now be written out and completed.

Continued on p. 216

Practical Exercise 10.3 *(Continued)*

1 When a low (logic 0) is applied to the lamp-test (LT) input, all segments are _____ [lit/not lit].
2 When a low is applied to the blanking (BI) input, all segments are _____ [lit/not lit].
3 In order to display decimal numbers 0 to 15, the blanking (BI) input must be held _____ [low/high].
4 When a low is applied to the ripple-blanking (RBI) input, all segments are _____ [lit/not lit], that is, the display is _____ [fully on/blanked].
This zero blanking action means that for multi-digit displays, a number such as 06 will be displayed as _____ [6/06].
5 Look in the catalogues and record the relevant details of the 7448 ic.

Questions

10.3 Which segments of the seven-segment display will be lit for each of the following decimal numbers?
(a) 1 (b) 4 (c) 7

10.4 The logic levels of the BCD decoder output are shown below. What will be the reading of the seven-segment display?
a b c d e f g
1 1 1 1 0 1 1

The seven-segment display with counter and latch

Practical Exercise 10.4 takes things a stage further by displaying the output from a 7490 decade counter. This is a 4-bit counter, using JK flipflops having separate divide-by-two and divide-by-ten circuits. Making the appropriate connections will achieve an overall count of 10. The opportunity will also be taken to incorporate a 7475 D-type latch.

Practical Exercise 10.4

To investigate the seven-segment display with counter and latch
For this exercise you will need the following components and equipment:

1 – 74LS90 ic (BCD decade counter)
1 – 74LS75 ic (latch)
1 – 7447A ic (BCD to seven-segment decoder)

1 – 0.3 inch seven-segment common anode display
7 – resistor (270 Ω)
1 – pulse generator (see Figure 1.5)
1 – +5 V DC power supply

Procedure (a)

1 Connect up the system from the block diagram shown in Figure 10.13. The pin connection diagram for the 7490 counter is given in Figure 10.14. Note that for a count of ten, pins 1 and 12 must be connected together.

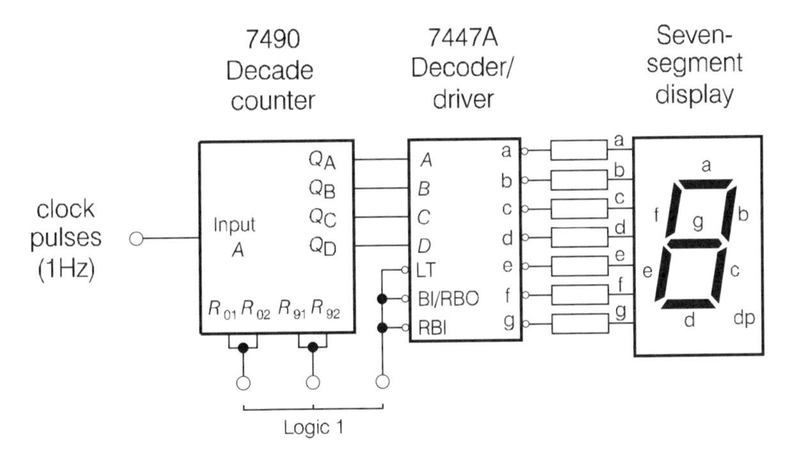

Figure 10.13 *The counter-display system: outline block diagram for Practical Exercise 10.4(a)*

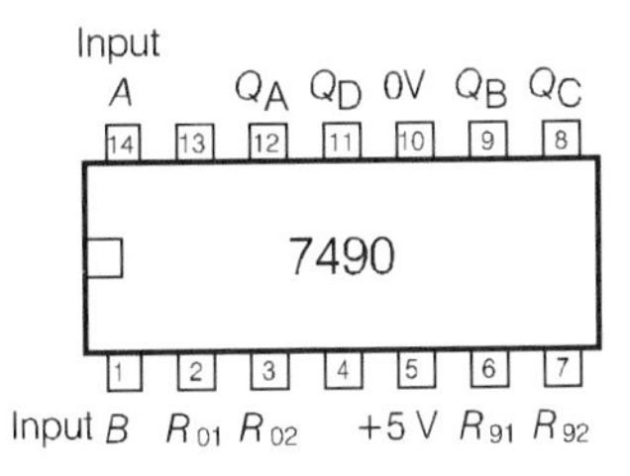

Figure 10.14 *The 7490 decade counter: pin connections*

2 Apply pulses to input *A* of the counter and observe the display. The counter should reset on the tenth pulse and the display then return to zero. Notice that for this arrangement the display follows the count.

Continued on p. 218

Practical Exercise 10.4 *(Continued)*

Procedure (b)

1 Modify the circuit according to the block diagram shown in Figure 10.15. The pin connection diagram for the 7475 ic is given in Figure 10.16

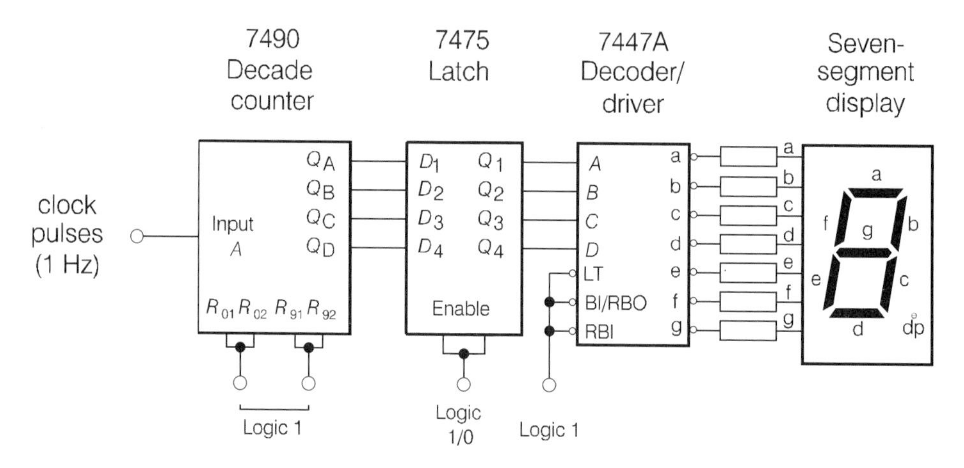

Figure 10.15 *The counter-display system: outline block diagram for Practical Exercise 10.4(b)*

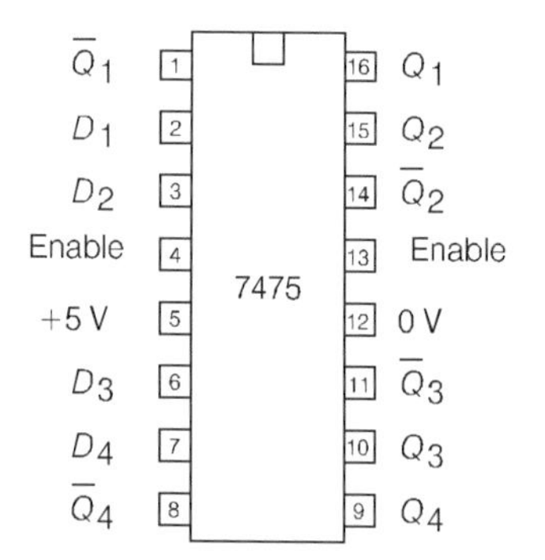

Figure 10.16 *The 7475 latch: pin connections*

2 Set the latch ENABLE input to logic 1 and apply a number of input pulses to the counter. Observe the display.
3 Set the ENABLE input to logic 0, continue the count and observe the display.
4 Return the ENABLE input to logic 1 and observe the display.

Conclusions

1. Give your observations on the behaviour of the display following the introduction of the latch.
2. What would be the advantage of using the latch for higher-frequency clock pulses?

Questions

10.5 Why is a decoder necessary in driving the seven-segment display from the BCD counter?

10.6 A system is required to count pulses up to a maximum of 999 using decade counters, decoder/drivers and seven-segment displays. Draw a suitable outline block diagram and show the logic levels present at the input and output of the decoder/driver for a count of 495.

The liquid crystal display (LCD)

In comparing the LCD with the LED, the following observations can be made:

(a) The LED generates light, whereas the LCD does not itself produce light, but controls it.
(b) The LCD has very low power consumption, making it ideal for portable (battery-operated) equipment.
(c) In dim light it is less easy to see an LCD display, where back lighting then becomes necessary.
(d) In good light conditions (sunlight) it is less easy to see an LED display.
(e) LCD is slower to respond (ms instead of μs).

The interface (namely the driving circuit) requirements are different from those for the LED. The LCD contains segments which will either remain transparent (corresponding to 'off') or opaque, that is, black ('on'). In order to cause a segment to show as black, a voltage has to be applied across that segment. Due mainly to the chemical nature and characteristics of the LCD, its life can be maximized if the energizing supply is alternating. A low frequency (40 Hz to 400 Hz) square wave from an astable multivibrator is found to be very suitable.

The dot matrix display

The general arrangement, showing a number of LEDs in columns and rows (the **matrix**) is given in Figure 10.17 and is known as a **five-by-seven display**.

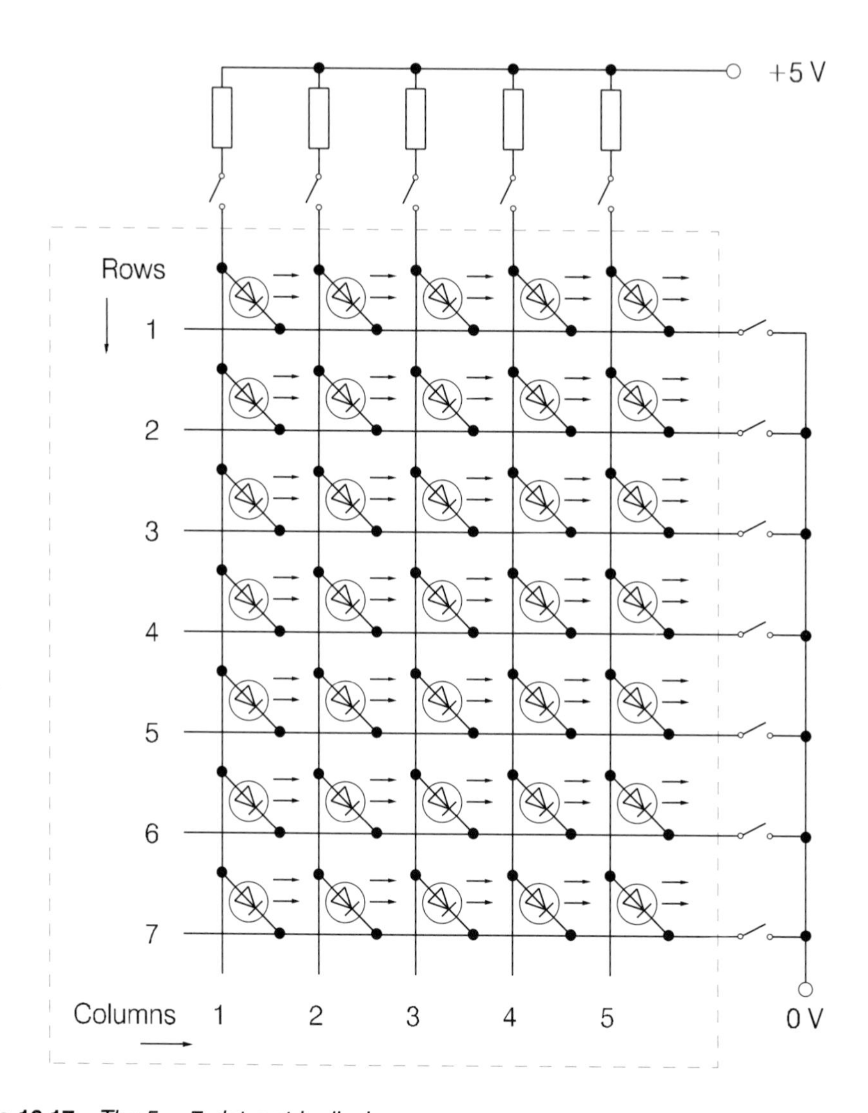

Figure 10.17 *The 5 × 7 dot matrix display*

To light a particular LED, its anode (connected to the column) must be supplied, through the series resistor, with +5 V. Its cathode (connected to the row) must go to 0 V. With the use of suitable switching, any number or letter can be displayed.

A number of units connected together and used in conjunction with a microprocessor driven controller, forms the basis of the **moving-message display**.

Finally, another LED arrangement called a **bargraph display** finds application in the analog world as a replacement for moving-coil meters and, in domestic audio equipment, as a sound level indicator.

A final word on LED devices

Listed in the catalogues and manufacturers' data sheets are so-called **$3\frac{1}{2}$digit** and **$4\frac{1}{2}$digit** displays. This is a reference to group displays which use a number of seven-segment devices in order to provide the required number of output digits.

Figure 10.18 *The $3\frac{1}{2}$ digit display*

Figure 10.18 shows a typical $3\frac{1}{2}$ digit display in which the left-hand digit (known as the $\frac{1}{2}$) will be blank or a 1 only. The remaining three digits behave as normal, and so, depending on the position of the decimal point, the unit will indicate up to 1999. It follows that a $4\frac{1}{2}$ digit display will read up to 19 999.

11 Analog and digital conversion

Many inputs are only available as analog signals (see the examples given in Chapter 1). Digital techniques are increasingly used for the signal-processing stage, with the resulting advantages of reduced noise and interference, increased speed of transmission and general improvement in performance and accuracy.

The two types of signal must therefore be correctly linked together, or **interfaced**. The need is thus for a device which will change analog signals into digital signals and the other way round (Figure 11.1). Since the signals themselves need to be electrical in nature, usually as a voltage rather than the position of a pointer, we will assume that the analog signal is already in this (electrical) state.

It will be appropriate, for a reason that should become clear later on, to deal first with the conversion from digital to analog.

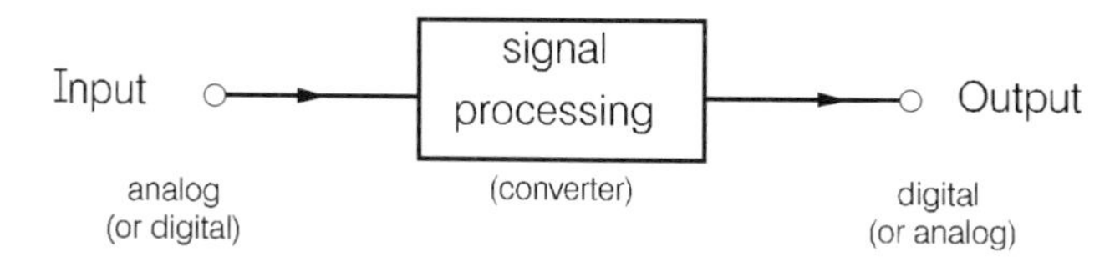

Figure 11.1 *The conversion process*

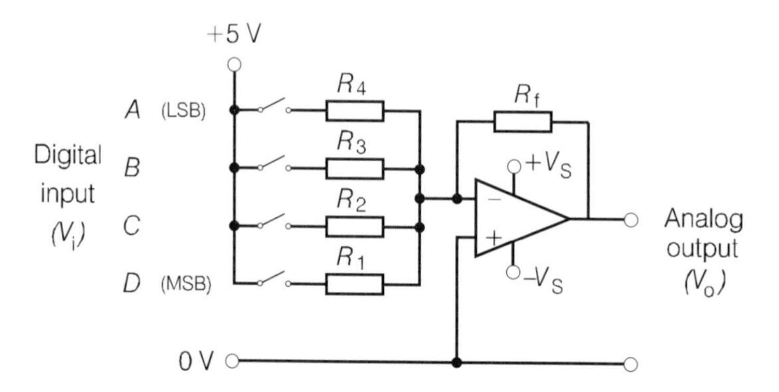

Figure 11.2 *The binary weighted DAC*

Digital-to-analog converters (DACs)

The binary weighted type

The diagram in Figure 11.2 shows the circuit of one type of DAC and uses an operational amplifier. The number of inputs to the amplifier is the same as the number of bits in the digital input.

Important points

- This particular example uses a 4-bit word, with A being the least significant bit (LSB) and D the most significant (MSB), as in a typical truth table.
- The value of R_1 to R_4 respectively is inversely proportional to the significance of the bit. For example,

  for input D (the MSB), $R_1 = R$

  for input A (the LSB), $R_4 = 8R$
- The voltage between the inverting (−) input and the 0 V line is virtually zero (within a few millivolts). This is the **virtual earth principle**. Hence the voltage across each weighted resistor when the appropriate switch is closed will be 5 V.
- The gain (V_o/V_i) of a particular input (say A) can be calculated from

  $V_o/V_i = -R_f/R_4$

  which for input A gives

  $V_o/V_i = -R_f/8R.$
- The gain will be largest for input D (MSB) where $R_1 = R$, and smallest for input A (LSB) where $R_4 = 8R$. The gain of each input is therefore weighted according to the value or significance of the bit.
- The (analog) output voltage V_o of the DAC, is given by

  $$V_o = -\left(\frac{R_f V_1}{R_1} + \frac{R_f V_2}{R_2} + \frac{R_f V_3}{R_3} + \frac{R_f V_4}{R_4}\right)$$
- The DAC behaves as a summing amplifier.

Before continuing with a practical exercise, let us see how a typical DAC may be set up. Look back at Figure 11.2 and remember that resistors R_1, R_2, R_3 and R_4 must be in the ratio of 1-2-4-8, respectively.

The feedback resistor R_f is chosen to provide a suitable maximum output voltage. Suppose therefore that, with all four digital inputs connected, the required analog output voltage is 7.5 V. The 4-bit input gives 16 levels including zero (a truth table with four inputs will have $2^4 = 16$ lines) and so there will be a total of 15 steps (a fence with 16 posts will have 15 gaps between the first and last post!). Each step must be equal to $\frac{1}{15}$ of 7.5 V, that is, 0.5 V.

Hence with only input A connected (the first step above zero) the output will be 0.5 V for an input of 5 V and so, using the formula for voltage gain,

$$V_o/V_i = -R_f/R_4$$

we get (disregarding the minus sign)

$$0.5 \text{ V}/5 \text{ V} = R_f/R_4 = 0.1$$

from which

$$R_f = 0.1R_4$$

or alternatively

$$R_4 = 10R_f$$

A reasonable (practical) value for R_4 would be 100 kΩ, thus giving

$R_3 = 50$ kΩ (preferred value 47 kΩ)
$R_2 = 25$ kΩ (22 kΩ)
$R_1 = 12.5$ kΩ (12 kΩ)
$R_f = 10$ kΩ (10 kΩ)

Practical Exercise 11.1

To investigate the DAC
For this exercise you will need the following components and equipment:

1 – set of resistors, values as immediately above
1 – 741 ic (operational amplifier)
1 – ±15 V DC supply
1 – +5 V DC supply
1 – DC voltmeter

Procedure

1. Connect up the circuit of Figure 11.2 using the ±15 V supply for V_s and the +5 V supply for the digital input. The pin connection diagram for the 741 ic is given in Figure 8.4.
2. Confirm the virtual earth principle by measuring the voltage between the non-inverting (–) input and the 0 V line. It should be within a few millivolts of zero. If it is not, then the 741 ic is probably faulty.
3. Measure the DC analog output voltage for the combination of inputs according to the 16-line truth table partly drawn below in Figure 11.3.
4. Draw up your own truth table and complete it using the measured results. The first three lines have been completed as a guide.

Digital inputs (V_i)	Calculated contribution to analog output	Measured analog output (V_0)
D C B A	D C B A	
0 0 0 0	0 V 0 V 0 V 0 V	0 V
0 0 0 1	0 V 0 V 0 V 0.5 V	0.5 V
0 0 1 0	0 V 0 V 1.0 V 0 V	1.0 V
etc.		

(continue and complete for a further 13 lines)

Figure 11.3 *Table of results for Practical Exercise 11.1*

5 Draw a graph similar to that shown in Figure 11.4. Note that the analog signal is negative-going since it is the output of an inverting amplifier.

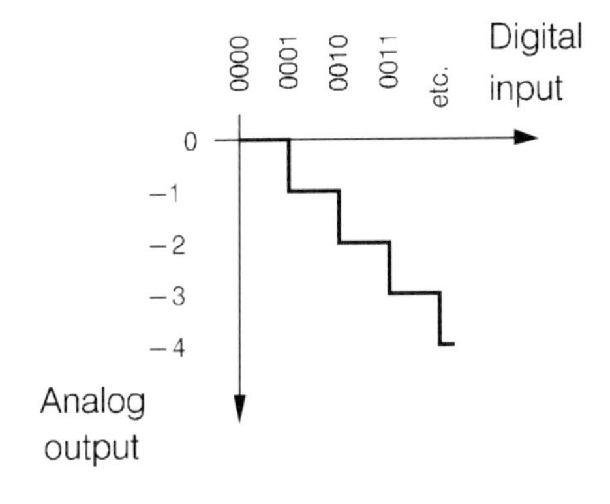

Figure 11.4 *Analog output versus digital input*

Important points

- The staircase shape of the graph in Figure 11.4 should show that there is a minimum change in the analog voltage that can be detected by the converter. This change, called **resolution**, has a value equal to the height of the step.

  For this exercise (a 4-bit DAC) the height of the step, and thus the resolution, was 0.5 V, which you may recall was the calculated value for the LSB.

Continued on p. 226

Important points (*Continued*)

- The resolution of an n-bit DAC is equal to 2^n, which when $n = 4$, would provide a resolution figure of 16. Put another way, the resolution can be given as one part in 2^n, that is, 1 part in 16 or 6.25%.
- The accuracy is given in terms of the number of steps in the digital output/analog input graph. In our example, there are 15 steps. Thus the accuracy is given as one part in 15, or one part in $(2^n - 1)$.

Questions

11.1 What do you think might be two possible problems of the binary weighted DAC?
(*Hint.* Consider resistor values and the resolution of the DAC).
Would these problems be greater or smaller if the DAC were a 16-bit type? Explain your answer.

11.2 A binary weighted 4-bit DAC uses an input voltage of 5 V for logic 1. The feedback resistor has a value of 10 kΩ and the output voltage from the least significant bit is to be 0.08 V.
Calculate the necessary values of the four input resistors and provide practical values from the E12 series.

The R-2R ladder network

This is the alternative to the binary weighted type of DAC – see Figure 11.5.

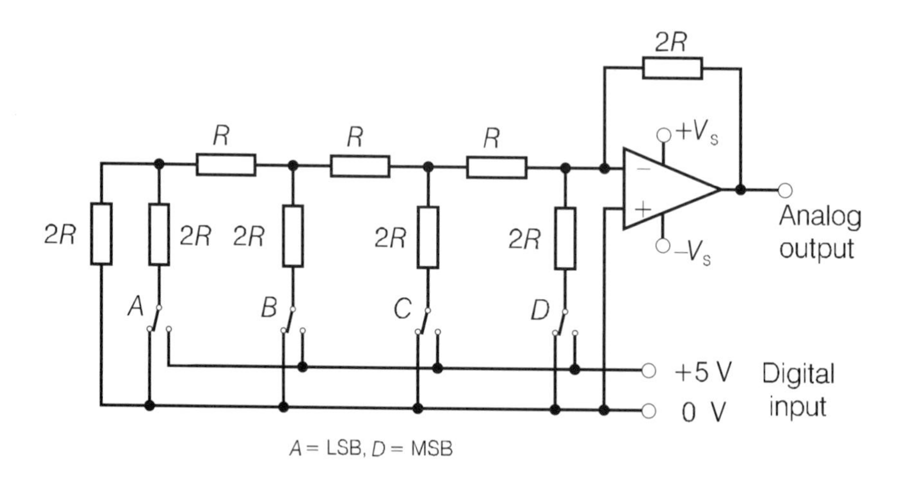

Figure 11.5 *The* R-2R *ladder network*

Important points

- The *R*-2*R* ladder network requires just two different resistor values, with one value being twice the other. The actual values are thus less important than the need to maintain the *R*-2*R* ratio.
- This type of DAC can be used with any number of bits and is available in integrated circuit form.

Questions

11.3 What advantage does the *R*-2*R* DAC have over the weighted type?

11.4 State the resolution of an 8-bit DAC

(a) as a whole number

(b) as a percentage.

11.5 Explain why an 8-bit DAC is likely to be more accurate than a 4-bit.

11.6 Look up and record the manufacturers' data on a typical integrated circuit DAC, such as the ZN428E.

Analog-to-digital converters (ADCs)

The counter-ramp type of ADC

The block diagram of a counter-ramp type ADC is shown in Figure 11.6.

Since the ADC has the need for a DAC, you will understand why it was necessary to deal with the DAC first of all.

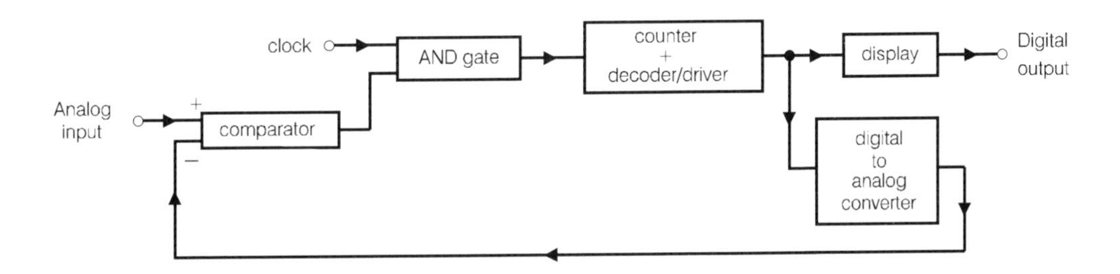

Figure 11.6 *A counter-ramp type ADC: outline diagram for Practical Exercise 11.2*

The action is as follows:

(a) The counter is reset to zero.
(b) Clock pulses are applied.

(c) Assuming that the AND gate is enabled, the counter then goes through its counting sequence.
(d) The output of the counter (which is digital) is taken to the input of the DAC and also to the display input.
(e) The output of the DAC increases in steps (as shown in Figure 11.4) and is applied to the inverting (−) input of the comparator amplifier.

Then, for the comparator:

(f) When the non-inverting (+) input is *more positive* than the inverting (−) input, the output will be positive (= logic 1).
(g) When the non-inverting (+) input is *more negative* than the inverting (−) input, the output will be negative (= logic 0).

Thus:

(h) When the (negative) DAC output is numerically less than the (positive) analog input, the AND gate is enabled and the count continues.

Conversely,

(i) When the DAC output is *just* (numerically) greater than the analog input, the AND gate is disabled and the count stops.

And so:

(j) The count stops when the digital output voltage (shown on the display) corresponds to the analog input voltage.

Practical Exercise 11.2

To make up and test an ADC
Refer to the outline block diagram shown in Figure 11.6. All the units, with the exception of the comparator, have been met in earlier chapters. A circuit for the comparator is given in Figure 11.7.

The brief

The analog-to-digital converter is to provide a digital output from an analog input of your choice between 1 V and 9 V. You are asked:

(a) To make up the complete unit. You will need to do some calculations for the DAC resistor values. Do not forget the (270 Ω) resistors for the seven-segment display!

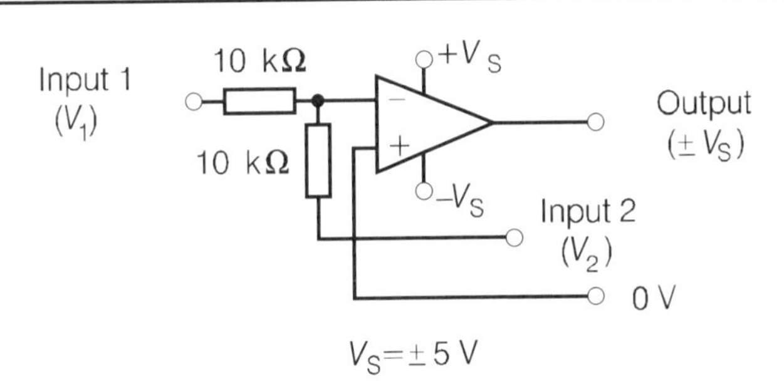

Figure 11.7 *The comparator: circuit for Practical Exercise 11.2*

(b) To test the complete unit (a suggested sequence of tests is given below).
(c) To provide a brief description of the theoretical operation of each individual unit and the unit as a whole, with comments on the actual performance of the overall system.

The test procedure

1. For the moment do not connect the output of the comparator to the AND gate.
2. Check that the AND gate operates according to its truth table.
3. Check the operation of the counter/decoder/seven-segment display by connecting the clock pulses directly to the clock input of the counter and omitting the AND gate.
4. If the counter is working correctly, the digital-to-analog converter can now be tested.

Important points

- The 7490 is a decade counter and will therefore automatically reset on the count of 10. (Hence the reason for making 9 V the upper limit for the analog input.)
- Remember to include the seven resistors in the display circuit. (Failure to do this will result in a burn-up of the LEDs!)
- The output of the DAC should be negative-going.

5. With the chosen positive analog voltage (V_1) and the negative DAC voltage (V_2) connected to the comparator through the two 10 kΩ resistors, the output from the comparator should be positive (approx. 5 V, i.e. logic 1) if V_1 is numerically *greater than* V_2.

   When V_1 and V_2 are *equal*, the output of the comparator will be zero.

Continued on p. 230

Practical Exercise 11.2 *(Continued)*

When V_2 becomes *greater than* V_1, the output of the comparator will be negative (approx. −5 V, i.e. logic 0).

6 Now make sure that all the connections are as in Figure 11.7 and check the operation of the unit as a whole.

Alternative types of ADC

The analog voltage used in Practical Exercise 11.2 was DC. Had it been a truly analog or changing voltage, for example 50 Hz sinusoidal, then the problem of satisfactory measurement would have been more noticeable.

Since it takes a finite time for the counter to count, the difficulties would obviously be more severe with large analog voltages requiring a high number of bits in the DAC, together with a high clock frequency. For example, for an 8-bit DAC, the binary output would be to eight places, which implies a count of 255.

The counter-ramp type of ADC is therefore really only suitable for small-magnitude, slow-changing analog signals. Two alternative types of ADC, both giving improved speed performance, are known as the 'tracking ADC' and the 'successive approximation ADC'. These are available in ic form as, for example, ZN435E and ZN427E respectively.

Questions

11.7 Explain briefly the essential differences between analog and digital signals.

11.8 Look up the manufacturers' data on the ZN425E integrated circuit to find:

(a) the voltage reference (full-scale output voltage)
(b) the number of bits

From this data, calculate the value of each voltage step.

11.9 Draw a fully labelled outline (block) diagram of an ADC set up to measure an analog voltage of 150 V.

12 Fault finding

Analog fault finding compared with digital fault finding

For analog systems the general measurement technique is firstly to check operating conditions by measuring the DC potentials at strategic points in the circuit. This will be achieved with the aid of a multimeter (either analog or digital type) and the voltage levels will be somewhere between two extreme values. This procedure is then followed by detailed examination of waveforms, a signal generator together with an oscilloscope being used to trace the signal through the system. For all these measurements a servicing manual, giving expected voltages and waveforms, is a boon!

Since with digital systems the logic levels are discrete, being either 1 or 0 (or changing between the two if pulsing) the system is somewhat easier to check. What has to be remembered is the range of allowable voltages corresponding to these logic levels, which of course depends upon the type of logic device being used (TTL, CMOS etc.).

As far as equipment is concerned, the multimeter can play a part in checking static (that is, stationary) levels and the oscilloscope, providing it has a reasonably wide bandwidth (10 MHz), is again useful for displaying pulse waveforms. There are, however, several items of equipment which can make a significant contribution to the problems likely to occur in typical digital systems. These will be dealt with in due course. It hardly needs saying that a pin-connection diagram for the various integrated circuits is an essential requirement!

Typical faults

We are normally concerned with the **output level** of a particular device and that is where the effort is usually concentrated.

Since the logic levels should be either 1 or 0 only, it follows that any fault must involve these levels being incorrect. The recognized maximum and minimum voltages for both TTL and CMOS are given in Figure 12.1. This table reminds us that the levels should be

(a) logic 0 (for TTL this means 0.8 V maximum), or
(b) logic 1 (for TTL, 2 V minimum), or
(c) pulsing (1 to 0 to 1 to 0 etc.).

What is *not allowed* is a level between 0.8 V and 2 V, which, by definition, is neither logic 0 nor logic 1.

Logic level	TTL (V_{CC} = +5 V) Voltage level	CMOS (V_{DD} = +3 V to +15 V) Voltage level
1	> 2.0 V	> 70% of V_{DD}
0	< 0.8 V	< 30% of V_{DD}
neither	between 0.8 V & 2.0 V	between 30% & 70% of V_{DD}

Figure 12.1 *Minimum and maximum voltage levels for TTL and CMOS*

The faults occur basically where the output of a gate is stuck (that is, not changing when it should) either at 1 or at 0.

Output stuck at 0 (at less than 0.8 V) where it should have gone to 1 (greater than 2 V)

The possible faults are:

(a) an ic internal s/c
(b) an open circuit power supply line to the ic (internally or externally)

Output stuck at 1 (greater than 2 V) where it should have gone to 0 (less than 0.8 V)

The possible faults are:

(a) an ic internal open- or short-circuit
(b) an open circuit 0 V line to the ic (internally or externally)

Output neither 1 nor 0 (between 0.8 V and 2 V). This represents a high impedance output state – replace the ic!

Most faults are due to open input and output circuits. Internal short-circuits in ic's will normally result in the ic becoming excessively hot.

The procedures

At the outset of any fault finding procedure it is helpful to be told (by the customer) of any known symptoms, e.g. intermittent operation, no output etc. Beyond that it is important initially to use one's senses to look, smell and feel.

Look for:

(a) signs of overheating (charring/discolouring of resistors)
(b) dry joints (even in equipment which has been working)
(c) breaks in printed circuit board tracks
(d) short-circuits between tracks (e.g. from solder)

Smell for overheated components (resistors).
Feel with finger (carefully):

(a) the surface of ic's
(b) resistors (for signs of excessive heat being dissipated)

Those with many years' experience (for whom this book is not intended!) will have a reservoir of knowledge which, coupled with either intuition or a good old-fashioned 'nose', will enable all sorts of short-circuit methods to be used. For those not so fortunate, the guiding motto has to be *'adopt a logical approach'*, with every action being, in turn, the result of the previous one.

Armed with the circuit and waveform diagrams, and pin connection diagrams, where do we start? Certainly the first thing to do is to check that the power supply is reaching the unit. You would not be the first person to discover, having spent some time getting nowhere, that the mains has not been switched on!

The idea of a logical approach consists of **tracing the logic path** through the system from the first stage to the last. If the fault is in the first stage, success is fairly immediate, but should it be in the later stages, then it will take longer, depending on the complexity of the system.

An alternative to this is the **half-split method**, which consists of dividing the number of units in the chain into two halves, taking either one or the other, and establishing if the fault is in that particular half.

If it is, then that half is divided into two sections and the fault condition identified as being in one or other of those sections, and so on.

If it is not, then the procedure is the same for the other half.

This may seem an over-complicated approach but, in theory, it should reduce the number of steps needed to locate the area in which the fault is present. In practice, since most systems do not conform to a simple chain-connected relationship, the half-split method needs a flexible approach.

Assuming now that we have located the fault to a particular area or sub-unit, what next? Still on the subject of power supplies, it is crucial to check that the ic itself is receiving the necessary voltage. This means measuring the voltage between the appropriate $+V_{CC}$ and 0 V pins of the ic's. The connection data is essential here, since not all ic's have the same pins for the incoming supply.

Remember that **for TTL**, $V_{CC} = 5\text{ V} \pm 0.25\text{ V}$, with 7 V an absolute maximum; **for CMOS,** $V_{DD} = 3\text{ V}$ to 15 V. It is wise not to remove or insert ic's with the power on.

The following fault finding examples assume that integrated circuits and printed circuit boards are used throughout. 'Track' may refer to copper strip board, which is commonly used for one-off units.

Some examples of fault conditions

These examples and the questions which follow are based on the work covered in previous chapters, but not necessarily, in every instance, with the identical unit.

There is no substitute for real live fault finding but, within the learning environment, a good appreciation can be acquired by putting on known faults and experiencing their effects. The reader can make up the following units to achieve this aim.

(a) The 2-input AND gate, shown in Figure 12.2

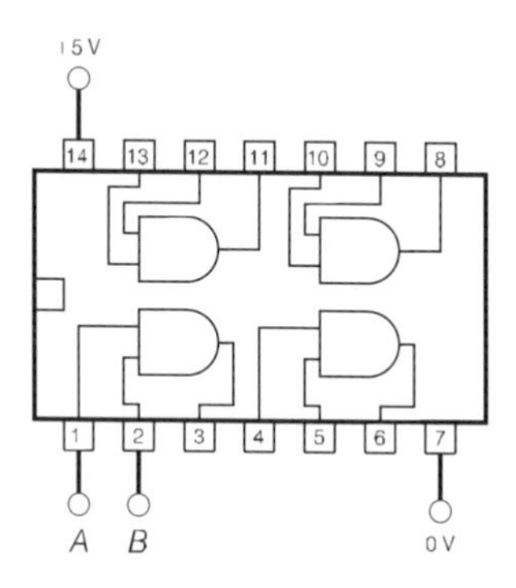

Figure 12.2 *The 2-input AND gate for examples (a) and (b)*

In this logical approach, no mention is made of the 'look, smell and feel' opening gambit, which must always be a prelude to the real business!

Symptoms	Output Z remains resolutely at logic 0 for all combinations of inputs.
Measurements	Supply to pcb = +5 V. Satisfactory. Voltage between pins 14 and 7 of ic = 0 V. Unsatisfactory.
Conclusions	Open circuit between positive supply input to pcb and pin 14, or in the 0 V rail.
Causes	Open circuit track on pcb (positive or negative rail). Pin 14 bent under or broken.

(b) The same AND gate (different fault!)

Symptoms	Output Z is 1 with input *A* at 1 and input *B* at 0. Otherwise gate behaves correctly. Pin 2 at 1 (even though input B at 0).
Measurements	Supply to pcb = +5 V. Satisfactory. Voltage between pins 14 and 7 of ic = +5 V. Satisfactory. Input *B* at 0 gives pin 2 at 1. Other lines in the truth table correct.

Conclusions	Open circuit between input *B* and pin 2. Pin 2 is floating and takes a logic 1 level.
Causes	Open circuit between input *B* and pin 2. Pin 2 bent under or broken.
Comments	If pin 2 was shorted to the +5 V rail rather than disconnected from input *B*, the second symptom above could not exist. It must therefore be an open circuit problem.

(c) The combinational logic circuit of Figure 12.3

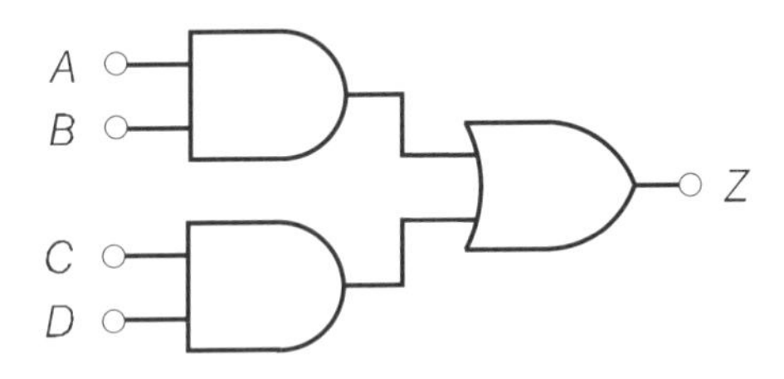

Figure 12.3 *The logic circuit for example (c)*

Symptoms	Output *Z* is permanently at 0.
Measurements	Supply to pins of all ic's = +5 V. Logic levels of both AND gates conform to their truth tables. Correct combination of logic levels at inputs to OR gate.
Conclusions	Fault lies between inputs of OR gate and output *Z*.
Causes	Faulty OR gate. Open-circuit track between output pin of OR gate and output *Z*.
Comments	In the case of a faulty logic gate, the only solution is a replacement.

Important point

- Without closer examination, further probing and measurement, or more details regarding logic levels, the causes must sometimes remain as alternatives. It is therefore only possible, in these instances, to give the *most likely* reason for the fault condition.

Questions

12.1 The combinational logic circuit shown in Figure 12.4(a) has the incorrect truth table given in Figure 12.4(b). Draw up the correct truth table and by deduction identify the possible cause of the fault.

Continued on p. 236

Questions (*Continued*)

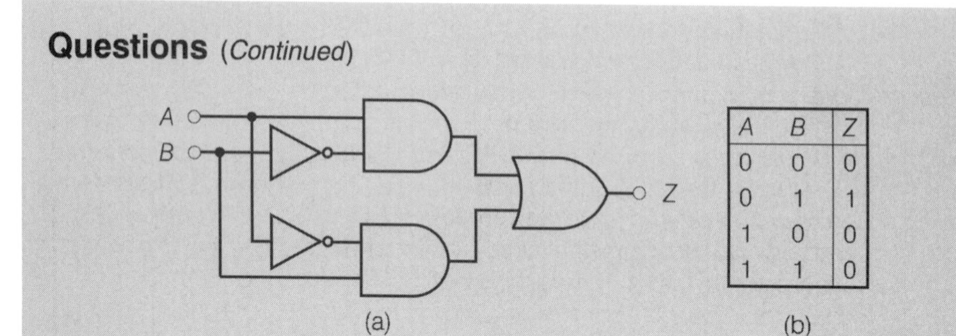

A	B	Z
0	0	0
0	1	1
1	0	0
1	1	0

Figure 12.4 *Logic diagram and truth table for Question 12.1*

12.2 The output of logic gate 1 in the circuit of Figure 12.5 is stuck at 1. Draw up the truth table and compare the output *Z* with and without this fault condition.

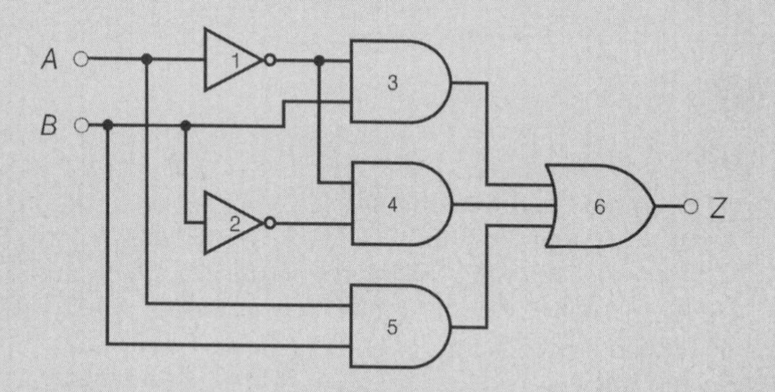

Figure 12.5 *Logic diagram for Question 12.2*

12.3 A counter using JK flipflops has the circuit shown in Figure 7.5. The counting sequence of the faulty unit is given in the table in Figure 12.6.

C	B	A
0	0	0
0	0	1
0	1	1
0	1	1
1	0	1
1	0	1
1	1	1
1	1	1

Figure 12.6 *Counting sequence for Question 12.3*

Symptoms	Output *A* is stuck at 1.
Measurements	Voltage between supply pins of all ic's = +5 V. Preset input pin levels for the flipflops are *A* = 0, *B* = 1, *C* = 1. Preset inputs on pcb are all 1.

Suggest, with reasoning, the possible cause of the fault.

12.4 A certain digital unit has the circuit of Figure 12.7. The normal working waveforms are shown in Figure 12.8.

Explain the effect of the following faults on the operation of the unit and draw the resulting output waveforms for each fault:

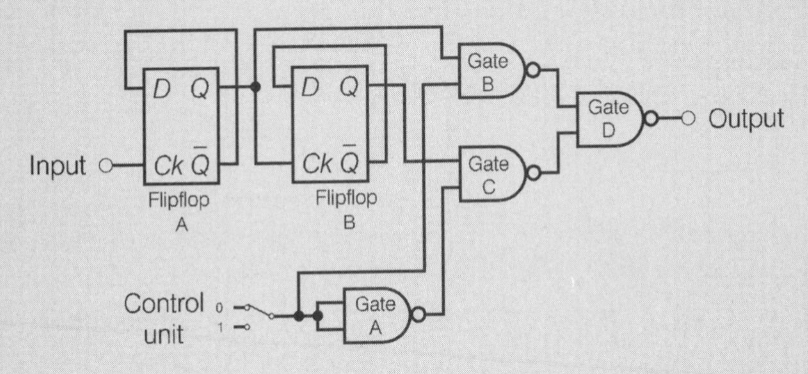

Figure 12.7 *The digital unit for Question 12.4*

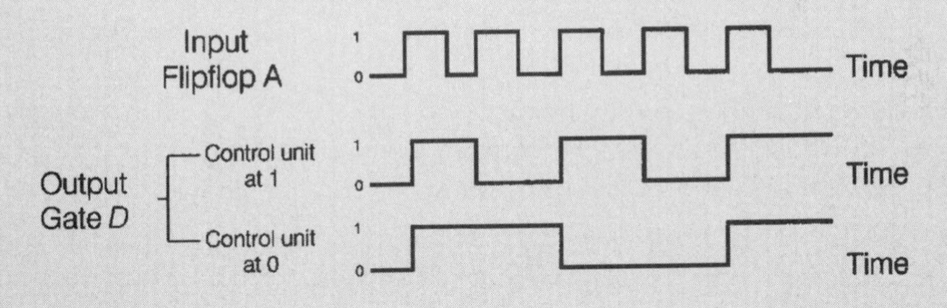

Figure 12.8 *Waveforms for Question 12.4*

1
(a) With control unit at 0 – output of gate *A* is stuck at 0.
(b) With control unit at 1 – output of gate *A* is stuck at 0.
(c) With control unit at 0 – output of gate *B* is stuck at 1.
(d) With control unit at 0 – output of gate *C* is stuck at 1.

12.5 A Schmitt trigger unit using NAND gates has the circuit shown in Figure 12.9. In this unit a slow astable multivibrator is used to turn a fast astable on and off. The correct waveforms are shown in Figure 12.10.

Continued on p. 238

Questions (*Continued*)

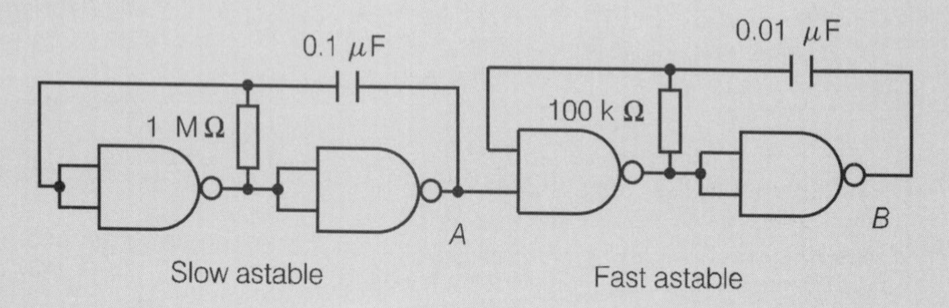

Figure 12.9 *The Schmitt trigger unit for Question 12.5*

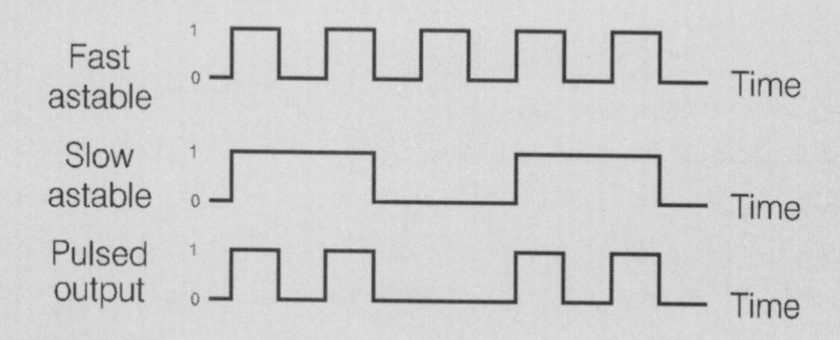

Figure 12.10 *Waveforms for Question 12.5*

Describe the effect on the operation of the unit of the following fault conditions:

(a) output *A* of slow astable stuck at 0.
(b) output *A* of slow astable stuck at 1.
(c) output *B* of fast astable stuck at 1.

12.6 An astable multivibrator has the circuit shown in Figure 8.11. State the likely effect on the output for each of the following component faults:

(a) R_1 open circuit.
(b) C_1 open circuit.
(c) Open circuit between junction of R_2/R_3 and the non-inverting input.
(d) R_2 open circuit.

12.7 A 555 timer monostable circuit is shown in Figure 8.18(a). State, with reasons, the effect on the output for each of the following fault conditions:

(a) Open-circuit track between 1 kΩ resistor and pin 2.
(b) R_1 very high (but not open-circuit).

12.8 The counter, decoder/driver and seven-segment display unit whose circuit is shown in Figure 10.13 has developed a fault.

Symptoms	Segment *g* does not light.
Measurements	Supply to all ic's = +5 V. With decoder LT input to 0, number 0 is displayed. Logic level check at decoder side of the resistors gives 0 at all points.

Describe what further tests should be applied and suggest possible causes for the fault.

12.9 A second counter unit has become faulty.

Symptoms	The counter output *C* is stuck at 0.
Measurements	These show that the decoder and display are working correctly.

Draw up the truth table and add the actual decimal display output to show the effect of this fault condition.

12.10 Refer to the block diagram of the analog-to-digital converter in Figure 11.6.

(a) Assuming all connections to be correct and all units powered up, state which unit or units could be faulty for the following symptoms:

(i) display reading goes from 0 to 9, resets and then repeats that sequence continuously

(ii) no display reading at all.

(b) Explain what you would expect the symptoms to be for each of the following component faults:

(i) open-circuit feedback resistor in DAC

(ii) open-circuit resistor in segment *e* of the display unit.

Equipment for digital fault finding

As mentioned previously, there are specific items of equipment for digital fault finding which will now be discussed.

The logic probe

This is a device which will indicate the presence of logic levels 1 and 0 with red and green LED indicators. Subject to it being possible to set the logic threshold level, the probe can be used for either TTL or CMOS ic's. It is usually powered from the circuit under test. (The logic levels and voltages for TTL and CMOS were given in Figure 12.1.)

The state of the LEDs and the interpretation of the logic levels are shown in Figure 12.11. An interesting and extremely useful aspect of this interpretation is that the probe not only gives the indication of logic levels 0 and 1 but also the indefinite state of neither 0 nor 1. This is important in fault finding since the indefinite state can

Red Green	Logic level	Voltage level
ON OFF	1	between 2.0 V and 5.0 V
OFF ON	0	between 0 V and 0.8 V
OFF OFF	neither 1 nor 0	between 0.8 V and 2.0 V
Flashing alternately	low-frequency pulse (less than 100 Hz)	changing
ON ON	higher-frequency pulse (greater than 100 Hz)	changing

Figure 12.11 *The logic probe: interpretation of the states of the LEDs*

cause a gate input to float between both levels. This situation is less readily appreciated with either an LED and resistor or a voltmeter arrangement.

The unit may contain a **pulse-stretching circuit** (our friend the monostable multivibrator!) to ensure that short-duration pulses (10 ns) last for long enough (100 ms) for the human eye to see the indicator.

The logic pulser (pulse generator)

The pulser forces a change of state at the input of a gate and allows the probe to be used to check the output. It is a **tristate output device**, meaning that the output can be switched into one of three states, namely logic 0, logic 1 and a high impedance state giving neither logic 0 nor logic 1. This latter state effectively isolates the pulser from the circuit when pulsing is not required. The forced change of input state referred to above requires a relatively large current (0.5 A being typical) which, if prolonged, could damage the overridden gate. The pulsewidth is thus limited to a value in the order of 1 μs at a frequency of 1 kHz.

In use, the pulser output tip is held against the ic pin, and a three-position switch can select single pulse, four pulse or a continuous pulse train. In spite of the large output current requirement, which can be either sourcing or sinking, the running current of the probe is usually only a matter of mA, thus enabling it to be powered from the circuit under test.

The logic monitor (logic clip, logic checker)

This device clips on to 14- or 16-way digital ic's and a series of LEDs displays the static or dynamic state of each ic pin simultaneously. State indication is given by LED off = logic 0, LED on = logic 1 and LED dim = pulsing. Like the previous device, it can be powered from the circuit under test. A typical application is for checking the operation of counters and shift registers.

The current tracer (current checker)

The tracer will indicate the magnitude and direction of the DC current flowing along a printed circuit track. It uses the magnetic effect of an electric current to detect the presence of short-circuits between tracks, without having to break into the track to insert a meter.

The logic analyser

This is a much more sophisticated piece of equipment, usually having 24, 32 or 48 input channels. Its use is principally for testing microprocessor systems, in which the logic levels on the data, address and control bus are sampled at certain instants in time and stored in memory for later recall and analysis. The cathode ray display will show all the data simultaneously, as waveforms or **timing diagrams**.

Answers to questions

Chapter 1

1.1 (a) analog
(b) digital
(c) analog
(d) analog
(e) analog
(f) digital
(g) analog

1.2 Easy to read with less likelihood of reading errors.

Chapter 2

2.1 See Figure A.1.

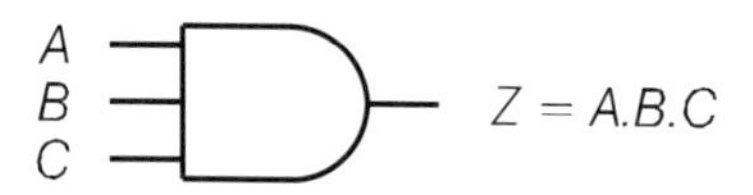

(a) The 3-input AND gate

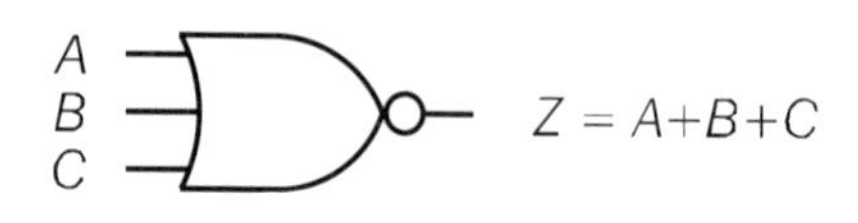

(b) The 3-input NOR gate

A	B	C	AND	NOR
0	0	0	0	1
0	0	1	0	0
0	1	0	0	0
0	1	1	0	0
1	0	0	0	0
1	0	1	0	0
1	1	0	0	0
1	1	1	1	0

Figure A.1 *Chapter 2, Question 2.1*

2.2 (a) Triple 3-input AND
(b) Dual 4-input NAND
(c) 8-input NAND
(d) Hex non-inverter

2.3 See Figure A.2.

A	B	$A.B$	$\overline{A.B}$	$\bar{A}$	$\bar{B}$	$\bar{A}.\bar{B}$
0	0	0	1	1	1	1
0	1	0	1	1	0	0
1	0	0	1	0	1	0
1	1	1	0	0	0	0

these are NOT identical

Figure A.2 *Chapter 2, Question 2.3*

2.4 See Figure A.3.

A	B	$A+B$	$\overline{A+B}$	$\bar{A}$	$\bar{B}$	$\bar{A}+\bar{B}$
0	0	0	1	1	1	1
0	1	1	0	1	0	1
1	0	1	0	0	1	1
1	1	1	0	0	0	0

these are NOT identical

Figure A.3 *Chapter 2, Question 2.4*

2.5 See Figure A.4.

A	B	$\bar{A}$	$\bar{B}$	$\bar{A}.B$	$A.\bar{B}$	$A.B$	$(\bar{A}.B+A.\bar{B}+A.B)$	$(A+B)$
0	0	1	1	0	0	0	0	0
0	1	1	0	1	0	0	1	1
1	0	0	1	0	1	0	1	1
1	1	0	0	0	0	1	1	1

these ARE identical

Figure A.4 *Chapter 2, Question 2.5*

2.6 '1'
2.7 Cost saving: the need for one type of gate only. Space saving: better utilization of gates within each ic leads to a likely reduction in overall number of ic's.
2.8 See Figure A.5

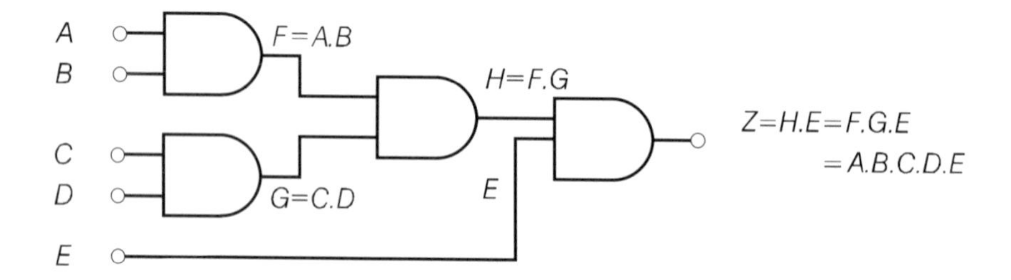

Figure A.5 *Chapter 2, Question 2.8*

2.9 See Figure A.6

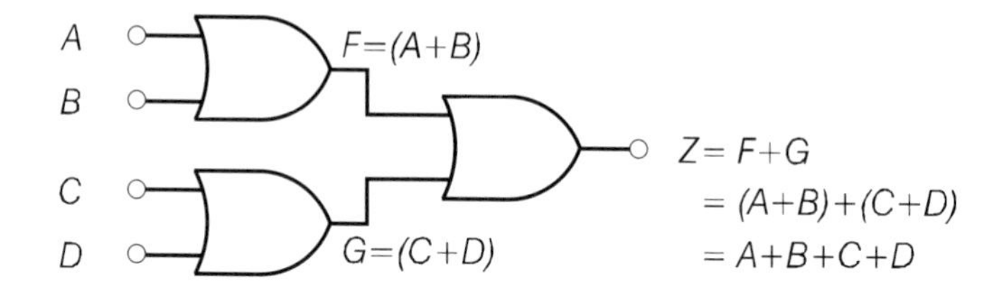

Figure A.6 *Chapter 2, Question 2.9*

2.10 See Figure A.7

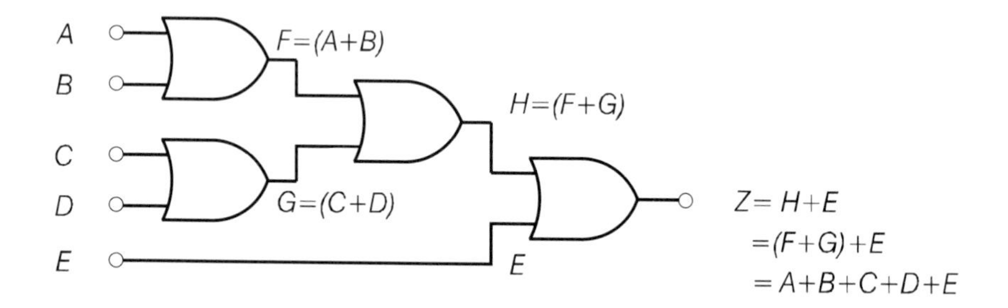

Figure A.7 *Chapter 2, Question 2.10*

2.11 See Figure A.8

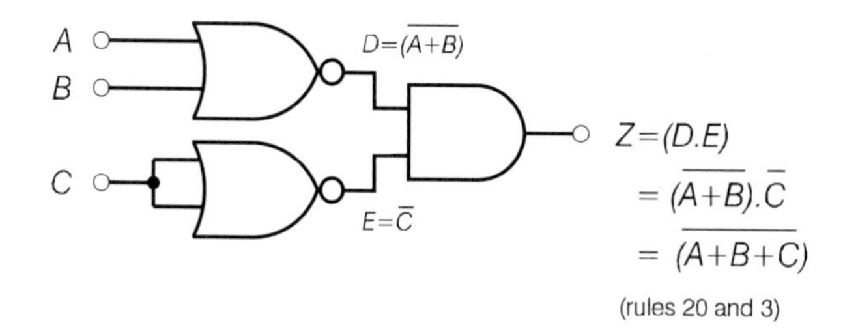

Figure A.8 *Chapter 2, Question 2.11*

2.12 See Figure A.9

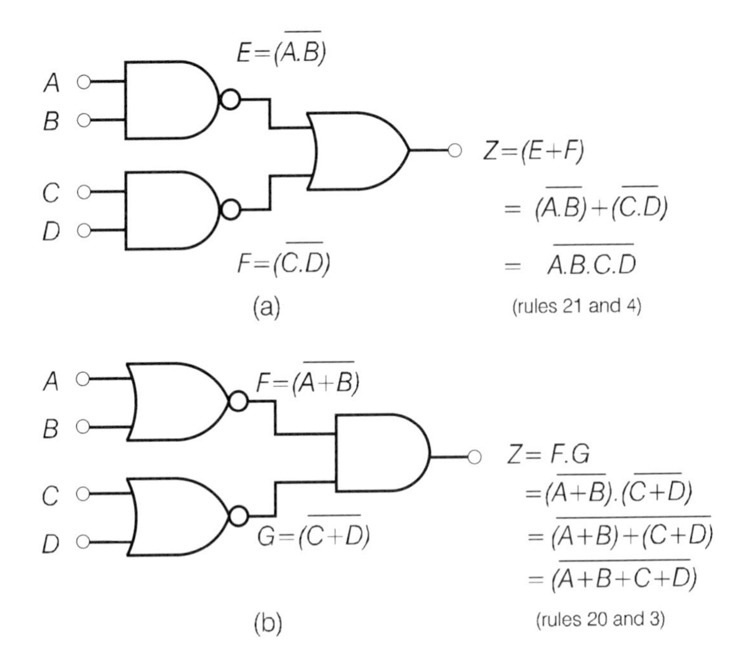

Figure A.9 *Chapter 2, Question 2.12*

2.13 See Figure A.10

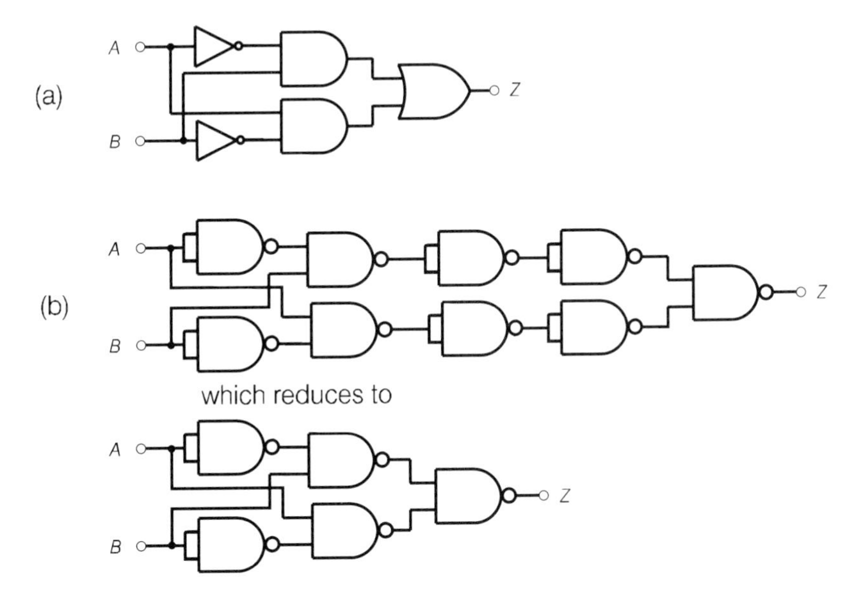

Figure A.10 *Chapter 2, Question 2.13*

2.14 See Figure A.11

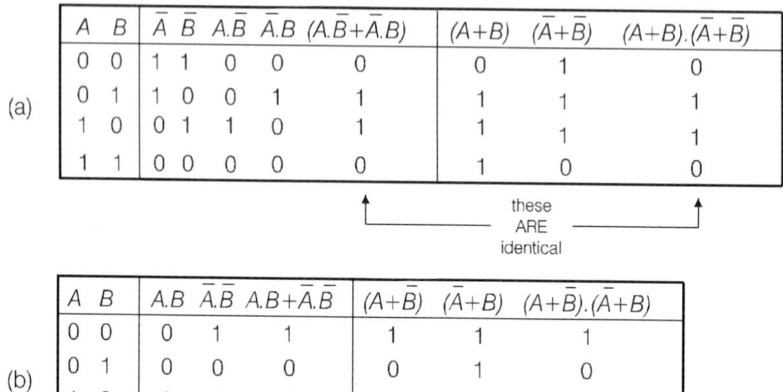

(a)

A	B	$\bar{A}$	$\bar{B}$	$A.\bar{B}$	$\bar{A}.B$	$(A.\bar{B}+\bar{A}.B)$	$(A+B)$	$(\bar{A}+\bar{B})$	$(A+B).(\bar{A}+\bar{B})$
0	0	1	1	0	0	0	0	1	0
0	1	1	0	0	1	1	1	1	1
1	0	0	1	1	0	1	1	1	1
1	1	0	0	0	0	0	1	0	0

(b)

A	B	$A.B$	$\bar{A}.\bar{B}$	$A.B+\bar{A}.\bar{B}$	$(A+\bar{B})$	$(\bar{A}+B)$	$(A+\bar{B}).(\bar{A}+B)$
0	0	0	1	1	1	1	1
0	1	0	0	0	0	1	0
1	0	0	0	0	1	0	0
1	1	1	0	1	1	1	1

Figure A.11 *Chapter 2, Question 2.14*

2.15 $Z = A \cdot B$
2.16 $Z = A$
2.17 $Z = B + A$
2.18 $Z = B + A \cdot C$
2.19 $Z = \overline{A + B + C}$
2.20 $Z = A \cdot \overline{B}$

Chapter 3

3.1 See Figure A.12.

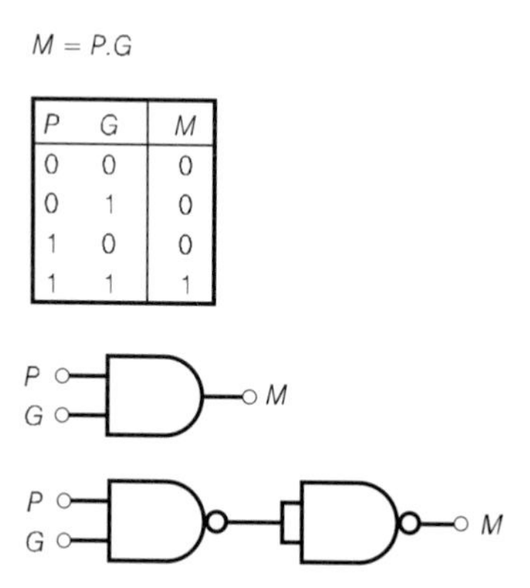

P	G	M
0	0	0
0	1	0
1	0	0
1	1	1

Figure A.12 *Chapter 3, Question 3.1*

3.2 See Figure A.13

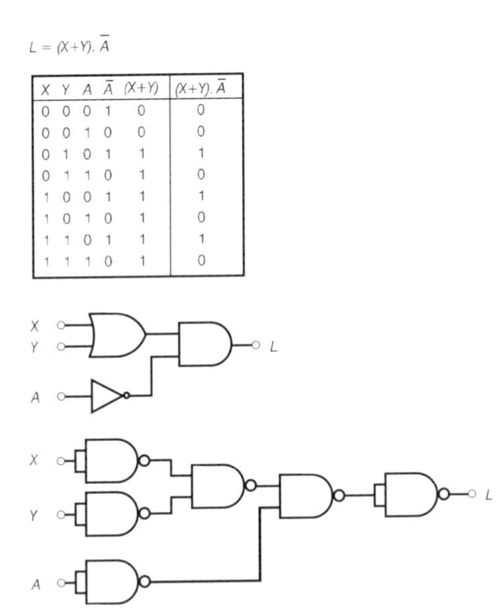
$L = (X+Y).\bar{A}$

X	Y	A	$\bar{A}$	(X+Y)	$(X+Y).\bar{A}$
0	0	0	1	0	0
0	0	1	0	0	0
0	1	0	1	1	1
0	1	1	0	1	0
1	0	0	1	1	1
1	0	1	0	1	0
1	1	0	1	1	1
1	1	1	0	1	0

Figure A.13 *Chapter 3, Question 3.2*

3.3 See Figure A.14.

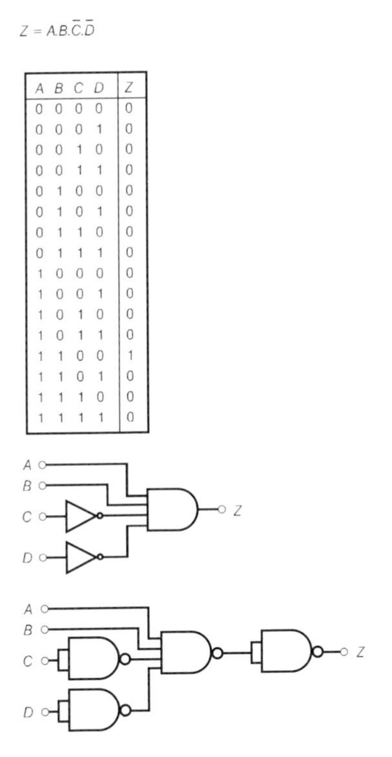
$Z = A.B.\bar{C}.\bar{D}$

A	B	C	D	Z
0	0	0	0	0
0	0	0	1	0
0	0	1	0	0
0	0	1	1	0
0	1	0	0	0
0	1	0	1	0
0	1	1	0	0
0	1	1	1	0
1	0	0	0	0
1	0	0	1	0
1	0	1	0	0
1	0	1	1	0
1	1	0	0	1
1	1	0	1	0
1	1	1	0	0
1	1	1	1	0

Figure A.14 *Chapter 3, Question 3.3*

3.4 (a) See Figure A.15.

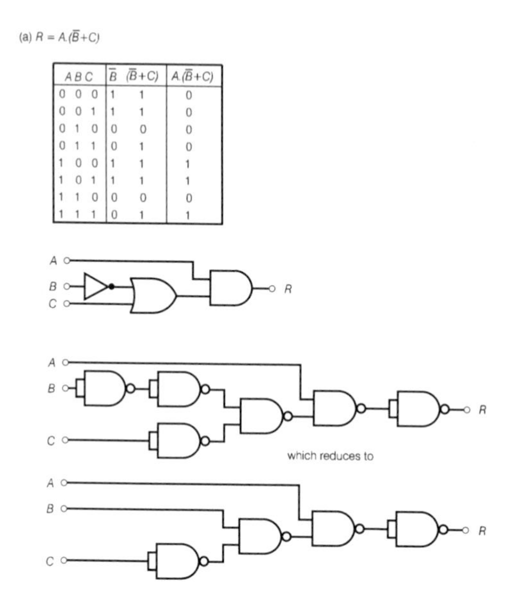
(a) $R = A.(\overline{B}+C)$

A B C	$\overline{B}$	$(\overline{B}+C)$	$A.(\overline{B}+C)$
0 0 0	1	1	0
0 0 1	1	1	0
0 1 0	0	0	0
0 1 1	0	1	0
1 0 0	1	1	1
1 0 1	1	1	1
1 1 0	0	0	0
1 1 1	0	1	1

Figure A.15 *Chapter 3, Question 3.4(a)*

(b) See Figure A.16.

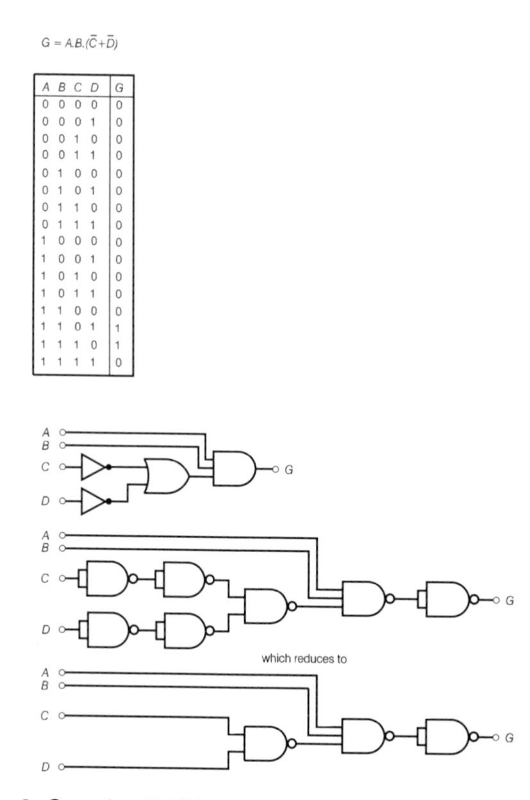
$G = A.B.(\overline{C}+\overline{D})$

A B C D	G
0 0 0 0	0
0 0 0 1	0
0 0 1 0	0
0 0 1 1	0
0 1 0 0	0
0 1 0 1	0
0 1 1 0	0
0 1 1 1	0
1 0 0 0	0
1 0 0 1	0
1 0 1 0	0
1 0 1 1	0
1 1 0 0	0
1 1 0 1	1
1 1 1 0	1
1 1 1 1	0

Figure A.16 *Chapter 3, Question 3.4(b)*

3.5 See Figure A.17.

$Z = A.\overline{B} + B.C$

A	B	C	$\overline{B}$	$A.\overline{B}$	$B.C$	$(A.\overline{B}+B.C)$
0	0	0	1	0	0	0
0	0	1	1	0	0	0
0	1	0	0	0	0	0
0	1	1	0	0	1	1
1	0	0	1	1	0	1
1	0	1	1	1	0	1
1	1	0	0	0	0	0
1	1	1	0	0	1	1

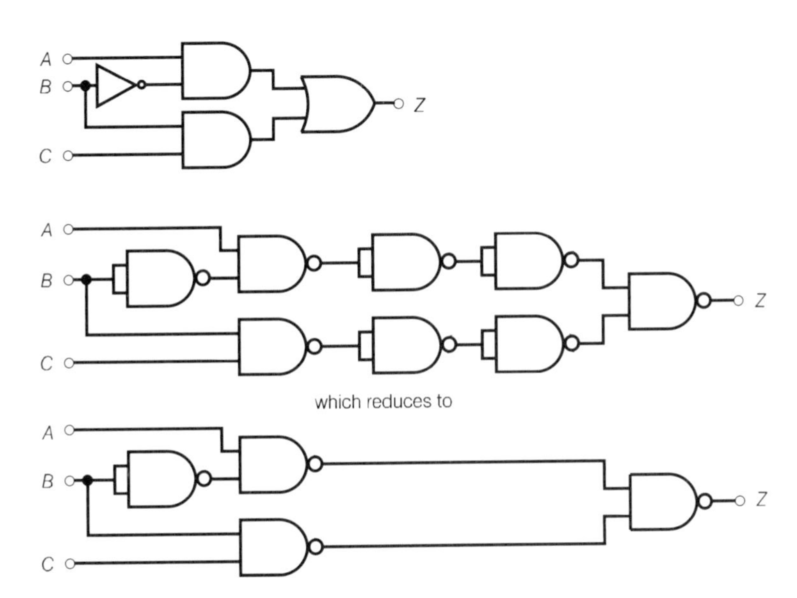

Figure A.17 *Chapter 3, Question 3.5*

3.6 See Figure A.18.

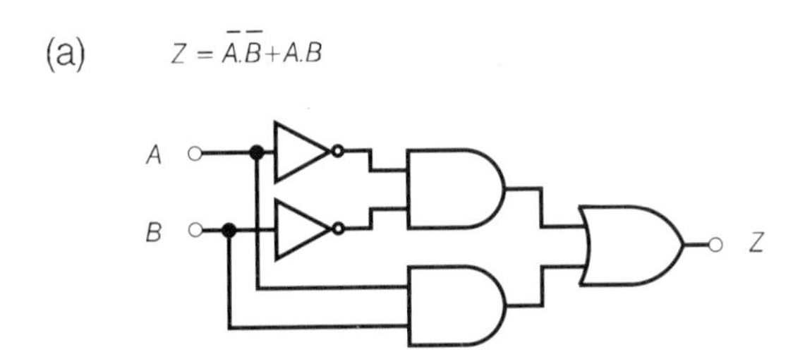

Figure A.18 *Chapter 3, Question 3.6*

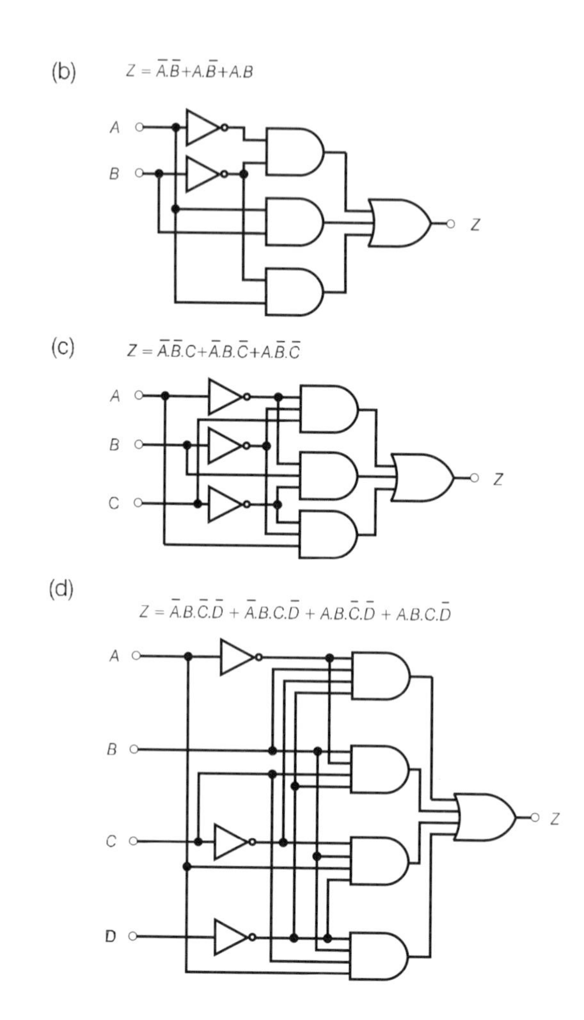

Figure A.18 *Continued*

3.7 See Figure A.19.

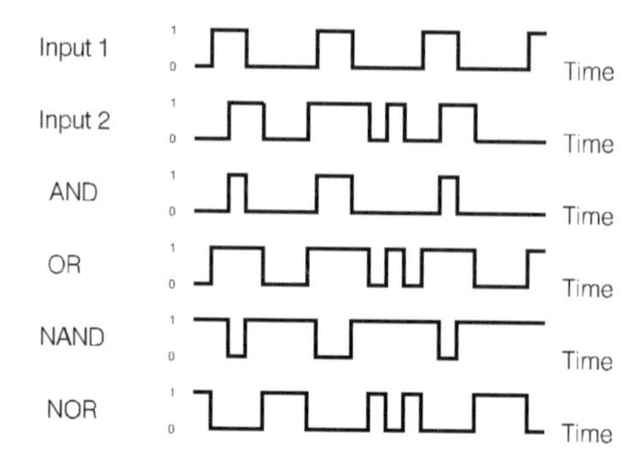

Figure A.19 *Chapter 3, Question 3.7*

3.8 See Figure A.20.

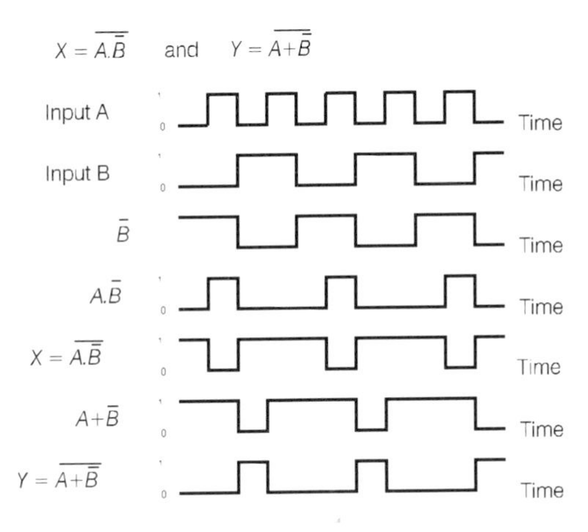

Figure A.20 *Chapter 3, Question 3.8*

Chapter 4

4.1 (a) $Z = A + B$
(b) $Z = \overline{A} + B$
(c) No simplification

4.2 (a) No simplification
(b) $Z = A \cdot B$
(c) $Z = A \cdot B + B \cdot C$
(d) $Z = A \cdot \overline{C} + B \cdot C$
(e) $Z = A \cdot B + \overline{B} \cdot C$
(f) $Z = A \cdot B \cdot C + \overline{B} \cdot \overline{C}$
(g) $Z = B$

4.3 (a) $Z = A \cdot \overline{B} \cdot \overline{C} + A \cdot B \cdot \overline{C}$
$Z = A \cdot \overline{C}$
(b) $Z = \overline{A} \cdot \overline{B} \cdot C + \overline{A} \cdot B \cdot \overline{C} + A \cdot \overline{B} \cdot \overline{C} + A \cdot B \cdot \overline{C}$
$Z = A \cdot \overline{C} + B \cdot \overline{C} + \overline{A} \cdot \overline{B} \cdot C$

4.4 $Z = \overline{A} \cdot B \cdot \overline{D} + \overline{B} \cdot D$

4.5 $Z = A \cdot \overline{B} \cdot C + B \cdot \overline{C}$

4.6 $Z = A \cdot \overline{B} \cdot \overline{C} + \overline{A} \cdot D$

4.7 $Z = A \cdot B \cdot C + \overline{A} \cdot \overline{B}$

4.8 $Z = \overline{A} \cdot D + \overline{B} \cdot C \cdot \overline{D} + \overline{B} \cdot C$
Additional term is $\overline{B} \cdot C$

4.9 $Z = \overline{A} \cdot B \cdot C + A \cdot \overline{B} \cdot C + A \cdot B \cdot \overline{C}$
No simplification
$Z = \overline{A} \cdot B \cdot C + A \cdot \overline{B} \cdot C + A \cdot B \cdot \overline{C} + A \cdot B \cdot C$
$Z = A \cdot B + B \cdot C + A \cdot C$

Chapter 5

5.1 (a) 3170 Ω, 1438 Ω
(b) 1969 Ω, 1438 Ω
5.2 3132 Ω
359 Ω

Chapter 6

6.1 (a) to (d) See Figure A.21

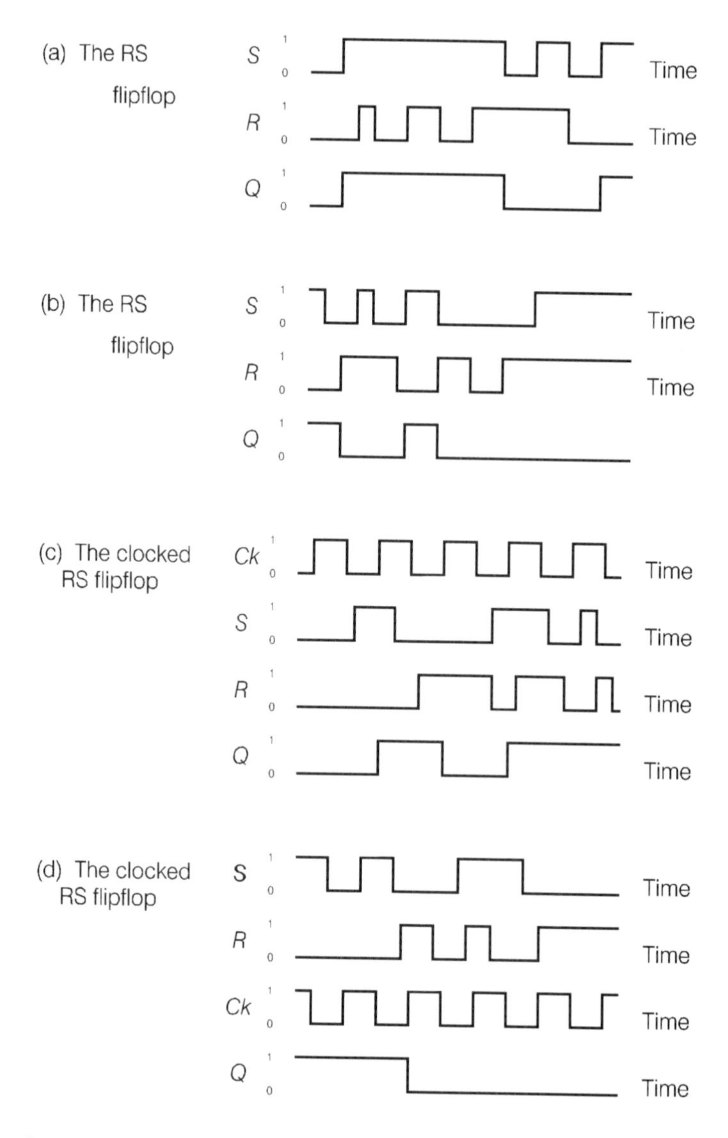

Figure A.21 *Chapter 6, Question 6.1*

6.5 (a) 0
(b) 1
(c) 1

6.7 D flipflop: data at *D* input is transferred to *Q* output when the clock pulse is applied. D latch: when the ENABLE (clock) input is held at logic 1, the *Q* output will follow the *D* input. With ENABLE at logic 0, the *Q* output will not respond to changes in *D* input.

Chapter 7

7.6 (a) 0101
(b) 0010
(c) 0001
(d) 0000

7.9 (a) 1 μs, 4
(b) 0.25 μs, 1

7.11 See Figure A.22.

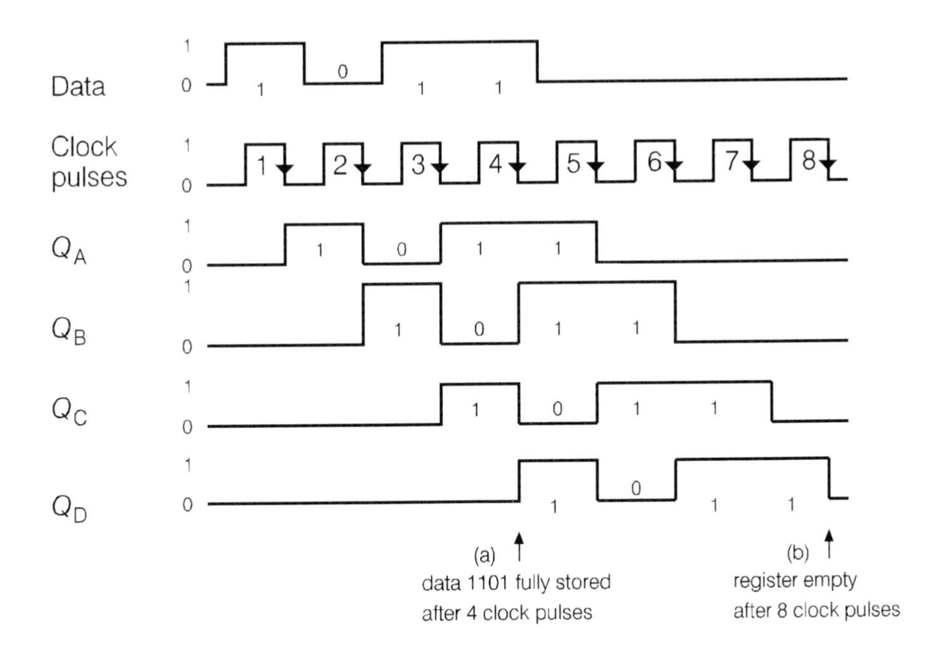

Figure A.22 *Chapter 7, Question 7.11*

7.12 See Figure A.23.

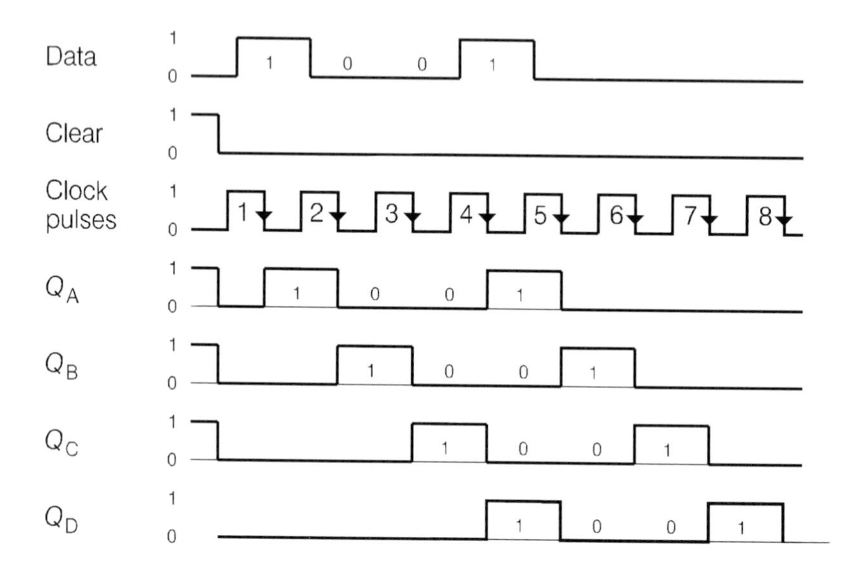

Figure A.23 *Chapter 7, Question 7.12*

7.13 Figure A.24.

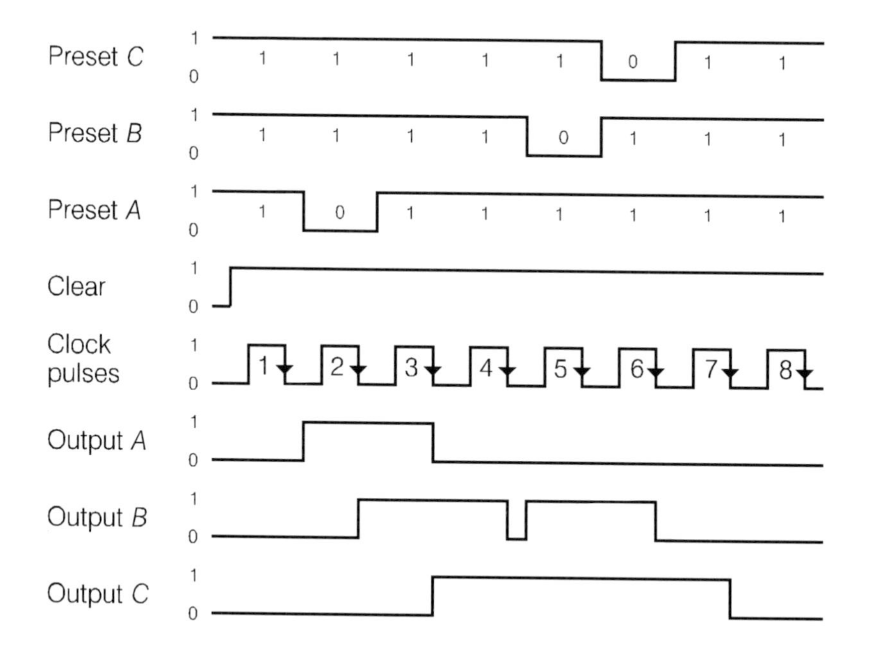

Figure A.24 *Chapter 7, Question 7.13*

Chapter 8

8.1 +1.18 V, −1.18 V, 2.36 V
8.2 $R_2 = 12\ \text{k}\Omega$, $R_1 = 1\ \text{k}\Omega$ would be suitable
8.3 R_1 decreases, hysteresis decreases
8.5 Mark–space ratio becomes 1:1
8.6 $(R_1 + R_2)$ can never be smaller than R_2
8.7 Lower frequency
8.8 (a) Connect pin 5 to variable DC supply
(b) Connect 2 diodes in inverse parallel, with associated series resistors, between pins 6 and 7

Chapter 9

9.1 (a) $Z = \overline{A} \cdot \overline{B} \cdot \overline{C} \cdot D_0 + \overline{A} \cdot \overline{B} \cdot C \cdot D_1 + \overline{A} \cdot B \cdot C \cdot D_3 + A \cdot B \cdot C \cdot D_7$
(b) $Z = \overline{A} \cdot \overline{B} \cdot \overline{C} \cdot D_0 + \overline{A} \cdot B \cdot \overline{C} \cdot D_2 + A \cdot B \cdot \overline{C} \cdot D_6 + A \cdot B \cdot C \cdot D_7$
9.3 (a) 16, (b) 16, (c) 24, (d) 32, (e) 220

Chapter 10

10.1 180 Ω (calc. 160 Ω)
10.2 700 Ω (use 680 Ω)
10.3 (a) b, c
(b) b, c, f, g
(c) a, b, c
10.4 9

Chapter 11

11.2 624 kΩ, 312 kΩ, 156 kΩ and 78 kΩ
11.4 256, 0.39%

Chapter 12

12.1 Faulty line: $AB = 10$
Cause: $A = 0$
$B = 1$ (floating input to NOT gate)
Possible fault: open circuit input to NOT gate

12.2

A	B	Z	Z (fault)
0	0	1	1
0	1	1	1
1	0	0	1
1	1	1	1

12.3 Open circuit input between preset *A* input pin and +5 V supply. Broken track or bent pin

12.4 (a) Output gate *D* at 0
(b) Output gate *D* at 1
(c) Output gate *D* identical to output flipflop *B*
(d) Output gate *D* at 0

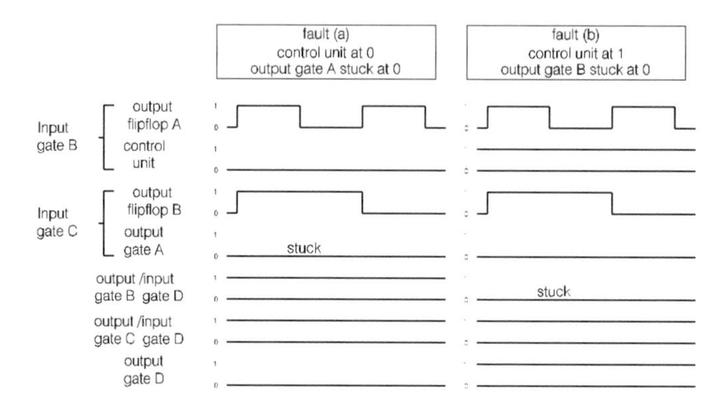

Figure A.25 *Chapter 12, Question 12.4 faults (a) and (b)*

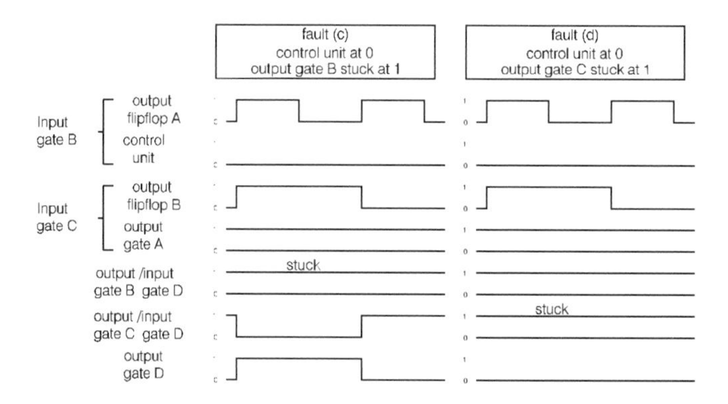

Figure A.26 *Chapter 12, Question 12.4 faults (c) and (d)*

12.5 (a) No output at either *A* or *B*
(b) Output *B* is continuous (fast astable)
(c) Output *B* is 1. Output *A* correct (slow astable)

12.6 (a) $+V_s$(sat)
(b) 0 V
(c) $-V_s$(sat)
(d) 0 V
N.B. No oscillations in each case!

12.7 (a) No trigger signal (pin 2 remains at 1). Output remains at 0
(b) Long charging time

12.8 Check logic level at right-hand side of bottom resistor:
If at 0, segment *g* is faulty
If at 1, resistor is faulty

12.9 Count (correct): 0 1 2 3 4 5 6 7 8 9 reset
Count (faulty): 0 1 2 3 0 1 2 3 8 9 reset

Index

Active low, 105, 111, 146, 177, 182, 213
Analog to digital conversion (ADC), 227–30
Analog signals, 2
AND gate, 7–9
ANSI, 8
Astable multivibrator, 165, 166–72
Asynchronous, 111, 119, 131

BCD code, 191
Binary addition, 198–206
  4-bit parallel, 201–3
  8-bit parallel, 203
Binary weighted DAC, 223–6
Bistable, 97, 165
Boolean Algebra, 7
BS 3939, 8

Channel addressing, 181
Charge storage, 66
Clock pulse, 4, 106–7
Clock pulse generator, circuit, 5
Clocked RS flipflop, 106–10
CMOS:
  NAND gate, 91
  NOR gate, 89–91
  NOT gate, 89
CMOS logic, 63, 88–91
Code converter, 197–8
Combinational logic, definition, 28
Comparator, 228–30
Contact bounce, 5
Converter, parallel-to-serial, 180
Counter:
  4-bit asynchronous, 134–5
  4-bit synchronous, 138, 139–43
  3-bit asynchronous, 129–33
  3-bit synchronous, 136–8
Counter-ramp ADC, 227–30
Counting down, 131, 133
Counting up, 131, 132–3
Counting sequence table, 128
Current tracer, 241

D flipflop, 110–19
Data selector, 181
Data transmission system, 188–90
De Morgan's theorems, 23, 24–5
Dead band, 158
De-bounced switch, circuits, 4
Decade, counter, 135, 216–19
Decoder, 194–8
Decoder/driver, 213–19
Delay flipflop, 110, 114
Demultiplexer, 186–90
Digital signals, 2
Digital to analog conversion (DAC), 223–7
Divide-by-two, 114, 122
Don't care/can't happen states, 53–6
Dot matrix display, 219–20
Duty cycle, 171

ECL logic, 63, 91–3
ECL OR/NOR gate, 91–3
Edge triggering, 106–7
Enable, 114, 123
Encoder, 190–4
Encoder, priority, 190
Exclusive OR and NOR gates, 12

Fan-in, 68–9
Fan-out, 68–9, 83–7
Fault conditions, some examples, 234–5
Fault finding:
  equipment, 239–41
  procedures, 232–3
Faults, typical, 231–2
Flipflop (Bistable), condition, 97, 165
Full-adder, 198, 200–6

Glitch, 58, 131

Half-adder, 198, 199–200
Hysteresis, 158, 164

Interfacing, 93, 222
  CMOS to TTL, 93–4
  TTL to CMOS, 94–5
  TTL to load, 88

JK flipflop, 119–23

Karnaugh map:
  four variables, 51–2
  minimal solution, 46
  three variables, 48–51
  two variables, 42–8
Karnaugh mapping, 40–61

Latch, 99–100
  D, 114–19, 216–19
  data, 114
  transparent, 114
LCD, 219–20
Least significant bit (LSB), 128, 223
LED, 5, 207–13, 219–21, 239–40
  as indicators, 210
Logic:
  analyser, 241
  expressions, simplification, 25–7
  families, table of characteristics, 73
  gate:
    packages, 14
    tristate, 87–8
    multiple input, 19–21
    summary, 13
  levels, 64–6, 71–2
  monitor, 240
  probe 239–40
  problem solving, 29–31
  pulser, 240
  rules and theorems, 23
  statement from truth table, derivation, 32–4
Lower threshold, 158
LSI, 63

Mark, 170
Mark–space ratio, 171
Master–slave JK flipflop, 123–6
Minimal solution, 46
Mode control, 153
Moduluo (modulus), 131
Monostable multivibrator, 165, 172–6
Most significant bit (MSB), 128, 223
MSI, 63

Multiplexer, 180–6, 188–90
  to implement a Boolean function, 184–6
Multivibrator, 164

NAND gate, 10–1
Negative edge triggering, 106–7
Noise, 70
Noise margin (noise immunity), 70–1
NOR gate, 11
NOT (Inverter) gate, 10

Open collector, gate, 78–80
Optoelectronics, 207
OR gate, 9
Oscillator, 164, 177
  crystal-controlled, 177–9

Parallel load, 150–2
Pin connection diagram:
  IN4148 diode, 69
  555 ic, 169
  7400 ic, 15
  7402 ic, 15
  7403 ic = 7400 ic, 15
  7404 ic, 15
  7408 ic, 15
  741 op amp, 160
  7410 ic, 32
  7411 ic, 118
  74121 ic, 175
  74139 ic, 187
  7414 ic, 163
  74147 ic, 192
  74153 ic, 182
  74161 ic, 139
  74192 ic, 141
  74194 ic, 153
  74283 ic, 205
  7432 ic, 15
  7442 ic, 197
  7447A ic, 214
  7474 ic, 112
  7475 ic, 218
  7476 ic, 120
  7490 ic, 217
  7493 ic, 135
  BC109 transistor, 76
  LED, 6
  seven-segment display ic, 213
PIPO, 150
PISO, 150
Positive edge triggering, 106–7
Power dissipation, 72

Propagation delay, 66–7
Pull-up resistor, 80, 85–7
Pulse length (width), 173, 174
Pulse repetition frequency (prf), 168–9
Pulse, input signal, 34–7

R-2R ladder network, 209–10
Race condition, 104
Race-round condition 123
Read, 115–16
Reset, 99–100, 101
Resolution, 225–6
Ring counter, 149
Ripple counter, 131
RS flipflop:
  unclocked, 101–6
  clocked, 106–10

Schmitt trigger, 158–64
Schottky diode, 95
Schottky TTL, 95–6
Sequential logic, definition, 97
Serial load, 144
Set, 99–100, 101
Seven-segment display, 210–19
Shift left, 152–9
Shift register (purpose), 143–4
  4-bit parallel load, 150–2
  4-bit serial load, 144–50
  recirculating, 149, 152
  universal, 152–5
Shift right, 145, 152–5
Sinking current, 82–5, 87, 88, 210
SIPO, 144
SISO, 147
Sourcing current, 82–5, 87, 88, 210
Space, 171
Speed-power product, 72–3
SSI, 63
Static hazards, 57–61
Stopping the count (resetting), 134
Successive approximation ADC, 230
Summary of logic gates, 13
Synchronous, 106, 119, 137

Time-division multiplexing, 180
Timer, 555 ic, terminal function, 176–7
Timing diagrams, importance, 104–5
Toggling, 121–2
Totem pole output, 75
Tracking ADC, 230
Traffic light system, 37–9, 56–7
Transient change, 58
Tristate logic, 87–8
TTL logic, 62, 73–87
TTL NAND gate:
  basic, 73–4, 76–7
  totem pole output, 74–5, 77–8
Two-transistor switch, 97–9
Typical faults, 231–2
Truth table, 8

Universal NAND and NOR gates, 17–18
Universal shift register, 152–5
Upper threshold, 158

Virtual earth, 223
VLSI, 63

Wired AND/OR gate, 80–2
Wired logic, 80
Write, 115–16